W9-BMZ-859

"A penetrating account of family life . . . [Finneran] brings the reader into an intimate understanding of the complicated ways in which families in general (and hers in particular) love one another."

— *USA Today*

"[Finneran] writes like a veteran and tells a story that is both sad and heartening . . . but ultimately triumphant."

— *Washington Post Book World*

"*The Tender Land* reminds us of how complicated, unique, and fragile an organism the family is . . . In this era of tell-all memoirs, it is a lovely surprise to find a memoir that celebrates family rather than what makes it fall apart." — *Boston Globe*

"There is an intimacy about Kathleen Finneran's first book that makes you feel you've met her before . . . She possesses a wonderful ability to write about everyday family life in a way that is always engaging." — *Fort Worth Star-Telegram*

"A rare and wondrous book."

— Tillie Olsen, author of *Tell Me a Riddle*

"Kathleen Finneran has pieced together a portrait of an ordinary family that has the contemplative beauty of lace: intricate and dense, translucent with love." — Brian Hall, author of *The Saskiad*

"Eloquent and moving." — *Advocate*

"Staggeringly good . . . Brilliant as much for what it withholds as for what it tells . . . Finneran is a master storyteller whose intelligence shines through her prose and whose prose shines with an incandescent, tightly controlled passion." — *Out*

"[Finneran's] parents and her siblings come vividly into view in the kind of stories people tell about one another around the kitchen table . . . A memento mori like no other." — *Booklist*, starred review

"Unforgettable in its restraint and quiet beauty."

— *Publishers Weekly*, starred review

The Tender Land

A Family Love Story

Kathleen Finneran

A MARINER BOOK
HOUGHTON MIFFLIN COMPANY
Boston • New York

First Mariner Books edition 2003

Copyright © 2000 by Kathleen Finneran
ALL RIGHTS RESERVED

For information about permission to reproduce selections from
this book, write to Permissions, Houghton Mifflin Company,
215 Park Avenue South, New York, New York 10003.

Visit our Web site: www.houghtonmifflinbooks.com.

Library of Congress Cataloging-in-Publication Data
Finneran, Kathleen.
The tender land : a family love story / Kathleen Finneran.
p.   cm.
ISBN 0-395-98495-5
ISBN 0-618-34074-2 (pbk.)
1. Suicide victims — Family relationships. 2. Teenagers
— Suicidal behavior. I. Title.
HV6546.F53 2000
362.28'3'092—dc21 [B] 99-089868

Printed in the United States of America

Book design by Victoria Hartman

QUM 10 9 8 7 6 5 4 3 2 1

For my mother and father,
in whose lives I see the evidence
of faith and love and labor

And for Sean

Set me as a seal upon your heart,
for love is stronger than death.

— *from the Song of Songs*

Acknowledgments

In writing this book, I was fortunate to receive assistance and encouragement from many people. I am grateful to David Gould and Kathryn Haslanger (who has guided and inspired me with her intelligence, goodness, and grace), of the United Hospital Fund, for allowing me a leave of absence so early in my employment and for their continued kindness and support; the MacDowell Colony for the Arts and Cottages at Hedgebrook for the space and time to write in such beautiful surroundings; Jan Figueira for her perpetual optimism and good cheer; Alene Hokenstad for her clear and compassionate thinking; Patricia McEntee for being helpful with the book's beginning; Wendy Surinsky for her honesty and enthusiasm and for the passion she has for her own work and for the work of those whom she admires; Julie Eakin for her ardent and intelligent reading of the book, in its many incarnations, and for the continuing dialogue she carries on with me about it; Laura Popenoe for reading and scrutinizing the manuscript with such care; Anne Barasch and Marlene Eskin for always asking to read more; Pat Dick for her early influence and sustained interest; Georgia Binnington for providing much-needed affirmation; Douglas Gaubatz for helping me to learn, through the example of his life and his work, how to look at things, and for teaching me that the adventure lies within the routine; Janis Irene Roddy for all that she has contributed in life and in friendship and in art; and Karin Cook for making me be more honest than I might have been and for sharing precious days that inspired the book's title piece.

I regret not being able to thank in person the two teachers who most influenced my writing. They are the late Sondra Stang, whose regard made me want to write well and whose support was instrumental in the publication of this book, and the late Stanley Elkin, who provided me with the recommendations I needed to advance from one place to the next, and whose work I looked to for instruction and inspiration.

I am deeply grateful to Anne Edelstein, my agent, whose presence in my life — personally and professionally — is one of the best things to have happened to me as a consequence of my writing this book; and to Elaine Pfefferblit, my editor, who saw some potential in the few pages I sent to her, waited patiently while I wrote the rest, and gave to the book, and to me, her unparalleled attention, intelligence, loyalty, wisdom, skill, and good counsel.

I thank my parents, Thomas and Lois Finneran, my brother, Michael Finneran, and my sisters, Mary Elder and Kelly Sonntag, for trusting me with the material of our lives and for always taking an interest in what I write. For enlarging our family and adding new life to it, I thank my sister-in-law, Sauni Van Pelt Finneran, and my brothers-in-law, Dan Elder and Duane Sonntag, and I thank, with the greatest love and delight, my nieces and nephews, Sarah, Jesse, and Allison Elder, and Stephanie and Nicholas Sonntag.

I thank, cherish, and am forever altered by the person most responsible for my having written this book — my brother Sean Patrick Finneran — and regret that he did not realize the joy of living longer.

Most especially, and most affectionately, I thank Roberta Swann — in whom my good fortune has its origins — for reading and refining what I wrote and for giving me the friendship I needed to begin this book and to finish it.

Contents

THE TENDER LAND

The Evidence of Angels

To those who have seen The Child, however dimly,
    however incredulously,
The Time Being is, in a sense, the most trying time
    of all.

— *W. H. Auden*

My mother believes she gave birth to an angel. She told me so when I stopped by one day for lunch, and though we have never discussed it, I imagine she told Michael, Mary, and Kelly just as matter-of-factly. "I think there was a reason he was only here for a short time," she said. "I think he was an angel sent to save someone."

My father was sitting across from me at the kitchen table. From merely looking at his face, I can usually tell exactly what he is thinking, especially if anything has been said that either of us might consider questionable. He has communicated silently with me since I was a child, staring at me from across a room or in the rearview mirror of the car until I look up to see what he wants to tell me. It is an unspoken language of astonishment, criticism, and condemnation. It has always kept us close.

The first time my father communicated with me this way I was five. He had picked me up from kindergarten. Usually my mother picked me up, but it was a beautiful fall day, and even though he was still in the construction business, and good weather was a commodity, my father was splendidly carefree sometimes, coming home early and taking us on long drives to undisclosed destinations, special places he wanted to show us.

But before we could go to wherever we were going that day, we had to drop off a boy in my class. His mother drove us to school and mine drove us home. When he saw that my father had come instead, the boy ran for the front seat, where I usually sat, so I climbed in back and sat behind my father. As he started the car, my father looked at me in the rearview mirror as if to say he recognized what the boy had done, usurping the seat that should have been mine. When we got to his house, the boy told my father to pull all the way up to the top of the driveway, as close to the front door as he could. "Closer. A little closer," the boy said. It was something my mother did every day without direction, the boy having instructed her the first time we took him home. He hated to walk any farther than he had to. Now the boy sat up high in the front seat to see out past the hood of the car, saying, "Just a few more feet." My father looked at me in the rearview mirror again. "Here is a real baby," his eyes said. I felt privileged then, and I didn't fight for the front seat later that day, as I usually did when we picked up Michael and Mary from North American Martyrs, the school I would go to the following year when I started first grade. Instead, I stayed in the back to watch in the rearview mirror for anything else my father might want to tell me.

It was almost twenty years later, and many words had passed unspoken between us by the time my mother revealed her belief that my younger brother, Sean, was an angel. It was a few weeks after Sean's death, and she spoke with such certainty and composure that I longed for my father to look at me and let me know what he was thinking. But he kept his eyes cast toward the table and continued to eat his sandwich without the slightest reaction, leaving me to wonder whether my mother's assessment of Sean's life and death was something he had already accepted, maybe even agreed with. He was unwilling to look at me, to meet my eyes in a way that might trivialize my mother's faith. Or perhaps the possibility of what she said consoled him, as it must have consoled my mother. Maybe the trauma of losing their fifteen-year-old son was lessened by believing his life was more than it might have been. Maybe faith has that effect.

My mother's faith has always been a natural, constant, almost practical part of our household. Her days begin and end in prayer. Each morning she sits in the living room with a large glass of instant iced tea and roams page by page through her prayer book, offering up her prayers for the living, her hopes for the dead. It is a time of privacy, but one she conducts in plain view, fielding her family's early morning inquiries calmly and quietly without ever looking up. When I still lived at home — as a child, as a teenager, and even as a young adult — I used to take my cereal into the living room, sit cross-legged on the couch across from my mother's chair, and eat my breakfast while she prayed. I never spoke and she never acknowledged me, until, having finished my cereal, I would get up to leave and she would hold her glass of tea toward me, asking if I'd mind adding more ice. It was a ritual. It was a way to participate, if only peripherally, in my mother's routine.

I don't have the same kind of faith as my mother, and as I sat there that day eating lunch with my parents, I turned her belief about Sean into something more like metaphor, though I knew that was not how she meant it. To her, Sean was not merely angelic; he was an actual angel. And I knew if I asked the obvious question — which of us was he sent here to save — she would have many answers. Maybe it wasn't just one of us. Maybe it was all of us. Or maybe it was someone we never even knew.

After we finished lunch, my mother got up and stood at the sink, staring out the kitchen window.

"Tom, the bird feeders are almost empty," she said to my father, and, turning to me, "We had a cardinal come this morning. I saw him sitting on the back fence when I woke up, and then he kept coming closer until he was right here on the windowsill. It's such a thrill to see that red in winter."

Above the kitchen window, a placard painted with flowers read, "What you are is God's gift to you. What you make of yourself is your gift to God." One of the many aphorisms that could be found hanging in our house, it was painted to look like a cross-stitch sampler and reminded me of the prayers my mother embroidered that hung above the bed Mary and I shared when we were little. One of the prayers — "Now I lay me down to sleep, I pray the Lord my soul to keep. If I should die before I wake, I pray the Lord my soul to take" — confused me. I didn't understand the word *keep* in terms of preservation. To me, it meant possession, permanent or otherwise. It meant asking my mother "Can we keep it?" whenever a stray animal wandered into our yard. It meant our neighbors keeping our goldfish while we were on vacation. Saying the prayer, I thought we were asking God to hold on to our souls — to keep them — while we slept, and I imagined God gathering them up every night and storing them somewhere, a large warehouse of souls being guarded until we got up again. And this is why I was confused: If God was already keeping our souls during the night, which we had prayed for him to do in the first place, it didn't make sense to ask him — if we died — to take what he already had. When I asked my mother about this, I wasn't able to explain my confusion clearly, and feeling frustrated by this inability, I kept my other questions to myself. How did God know what time we were going to wake up? I wondered. Did our souls come back automatically as soon as our eyes opened? What if my soul got mixed up with Mary's? Sometimes I woke up on her side of the bed and she woke up on mine, with no memory of how it happened. Did God have a system to keep track of such stuff?

As a child, saying that prayer every night, lying in bed below the sampler my mother had stitched, I never considered the possibility that any of us would die in our sleep. Just as I never thought it would happen when, if Michael, Mary, and I had been fighting, my mother made us apologize before we went to bed, telling us we would feel bad forever if one of us died during the night and we never got the chance to say we were sorry. But now it had happened, and I knew, too well, what my mother meant. Sean hadn't died in his sleep, but his death was sudden. None of us thought one day that he would not be here the next. And though we had no quarrels with him that had gone unforgiven, it didn't matter. He had killed himself. For the rest of us, there could be no greater guilt. We had not seen his pain, and for that we would always be sorry.

My father went outside to fill the bird feeders. Watching him, my mother tapped on the window and pointed toward the fence. The cardinal had come back. "Come see," she told me. The cardinal flew closer to my father and followed him as he finished filling the feeders. It was the food, of course, that the cardinal was following, but when my father came back into the house, the cardinal, instead of perching on one of the feeders, sat on the empty birdbath and stared at the kitchen window as if it were waiting for someone to come out again, and then it flew up and stood on the windowsill, as it had when my mother saw it that morning, and looked at us through the glass.

"Hi, pretty bird," my mother cooed. "Hi, pretty boy." We had been watching the cardinal for only a few minutes when Kelly came home. The youngest of us, she was twelve and still in grade school when Sean died. I was twenty-four, Mary and Michael two and four years older.

Kelly threw her coat on a chair and her books on the table. "What are you looking at?" she asked.

"A cardinal," my mother answered.

"What's the big deal about a cardinal?" Kelly went to the refrigerator and got out the milk and then pushed herself between us at the window. She was the only child now of what my mother referred to as her second family, Sean and Kelly born so many years after Michael, Mary, and me. She looked at the cardinal, then turned to my mother. "Don't even try to say that's Sean," she said, and seeing a smile on my mother's face, my father and I started laughing.

"I mean it," Kelly said. She was blunt about everything, including my mother's beliefs, and I imagined her rolling her eyes at the idea of Sean as an angel. "Yeah, right," she'd say, ready to tell us all the ways he wasn't.

When my mother went out to sprinkle some seeds on the

windowsill, I thought the cardinal would fly away, but it didn't. My mother said something to it and then she came back in and stood at the kitchen sink again, watching it through the window. "What's wrong, little guy?" she asked. "Aren't you hungry?" The cardinal looked at her for a few minutes and then flew off to the telephone wire, the tree, and out of the yard altogether. "Goodbye, little guy," my mother said. "Goodbye, pretty red bird."

As I stood there with her, watching nothing now, I thought about how much she and Sean sounded like each other. They both talked easily and openly to animals, using the same tone of voice, sometimes even the same words. "Goodbye, little guy," my mother called out to the cardinal. "Go on, little guy, you're free now," I had once heard Sean say to a frog. We had been riding our bikes on the river road that runs along the Illinois side of the Mississippi, just north of where we lived in the suburbs of St. Louis. It was a Saturday near the end of October, a few weeks after Sean's fifteenth birthday, and we had planned a longer ride than the one we usually took to the Brussels Ferry and back. This time, instead of touching the ferry sign and turning around, we would board the Brussels Ferry with our bikes, ride up the other side of the river to a ferry farther north, cross, and come back down. Sean hoped to reach the town of Hamburg. "Brussels and Hamburg in the same day," he said. It was his dream to ride to all the towns in Missouri and Illinois with European names. Florence, Rome, and Athens. Frankfurt, Strasbourg, Vienna, Versailles.

"We'll pass through Batchtown and Nutwood, too," he told me. He had drawn a map and slipped it into the plastic sleeve of his handlebar bag. Batchtown and Nutwood meant as much to him as Brussels and Hamburg. It was the names of places that he loved.

After the first few miles, Sean's map was already unreadable. It was the same with every map he made, drawn meticulously and sized to slip into the special handlebar bag he had bought to hold his maps in place so that he could read them while he rode. We never got very far before they were obscured by things he saw and stopped for, rocks and wildflowers mostly, leaves and weeds and sometimes money. This time it was two giant fern fronds full of spores and some tiny orange flowers that were blooming beside them. He planned to scrape the spores off the ferns and look at them under his microscope. The flowers? They were pretty.

"Did you know that people used to think that carrying fern spores could make you invisible?" he said.

We were passing all our favorite places — the house with the word PIES painted on the porch rail, the fish-fry stand where we always stopped for soda. We were on a mission: Hamburg or bust.

"When was that?" I asked.

"I can't remember. The Middle Ages maybe. I read it somewhere."

"You mean like they'd put the spores in their pocket or something and then think they were invisible? Couldn't they see themselves? Even if there weren't mirrors, they could still see their bodies."

"Maybe they became invisible to other people but not to themselves."

"Either way, it doesn't make much sense."

"Your gears are slipping," he said.

"Only the low ones."

"How can you stand riding that way?" he wondered.

When we reached the small park where we always stopped for lunch, we walked our bikes across the grass to a picnic table that stood beneath a tree beside the river.

"Table, tree, trash can," Sean said. "This would be a good place to teach Sarah the letter *t*."

"You're teaching Sarah the alphabet already?"

"No, but someday I will be," he said. Mary's daughter, Sarah, was four months old and not much time went by that Sean wasn't talking about her. He took his unclehood seriously, riding his bike to Mary's nearly every day to see her and supplying us with daily updates on what she was doing. "Table, tree, trash can" was the kind of thing he said a lot during those days, as if he had altered the way he experienced the world, or his expression of it, to meet the needs of his newborn niece. One of the things she needed most, he decided, was to know the name of everything she encountered. "School bus, Sarah," he would say as he pushed her in her stroller. "Car, Sarah. Stop sign. Sprinkler, Sarah. Kitty. C'mere, kitty," he would say. We had all grown used to his stopping midsentence to name something whenever she was with us. "Home, Sarah," I heard him say once when they returned from a walk. "Home, Sarah," he whispered as he lifted her, sleeping, out of her stroller.

He was already looking forward to the time when she would talk. "What do you want to tell me?" he would ask her, and she would kick her legs a little, fix her eyes on him, and smile. "Do you want to tell me about your duckie?" he would ask, waving it in front of her face, and he would rock and talk, taking both their parts, asking her questions and answering them for her.

"I wonder what her first word will be," he would say sometimes, but he would be dead before she said it. "Door," she would say one day, watching us walk through it. "Door," she would tell us again as we kissed her goodbye.

Table, tree, trash can. It was a spare assessment of the surroundings, but it was accurate. There was not much else around.

"River, clouds, sky," Sean said, and he looked at me and grinned.

"Boy, bike, bird. Snap out of it," I said.

He squinted toward the sky, then touched my arm. "Where's the bird?" he asked.

When we reached the picnic table, he threw down his bike. "Oh no! Oh God!" he screamed, and he started to cry.

Near the center of the table, a frog was stuck in a pink mound of bubble gum. If it had struggled to free itself, it had given up, and it sat there panting, its body expanding and contracting so fiercely it looked as if it would soon explode. A brown river frog, it had turned gray.

"Somebody did that to him," Sean cried. He took a cup out of his bag. "Go get some water from the river," he said, and when I returned, he was stroking the frog's back with his finger as he worked his pocketknife under the gum. He stuck his finger in the water and then ran it, wet, over the frog's back. "Do that," he told me, and then he began digging deep below the wad of gum, almost into the wood of the table, to keep from cutting the frog. He had stopped crying, but his eyes and face were wet with tears.

"Maybe he just jumped into it," I said.

"No. Somebody *did* it to him. Some fucking asshole," he said. I had never heard him talk like that.

"How do you know?"

"Because frogs have strong legs. If he jumped on this, he could jump out of it. He might take some gum with him, but he wouldn't get stuck."

He paused for a moment to wipe his face on his sleeve, and then he pointed his knife at an indentation where the frog's front feet were stuck. "See?" he said. The same mark — the size of a thumbprint — encircled the back feet. "Someone held him down. Real funny," he said.

He freed the wad of gum from the table and lifted the frog. The gum looked like a small pink pond beneath the frog's body. It reminded me of a ceramic figurine my mother had on her dresser of a little bird swimming on a puddle of blue porcelain.

The frog seemed less frightened. Its heart was beating more slowly and its color was turning back to brown. We sat down on the picnic table, and as I continued stroking the frog, Sean gently removed the gum from its feet and legs and belly. "Poor frog," he said, in a voice like my mother's. "Poor little frog."

When he was finished, he carried the frog to the bank of

the river and set it on a spot of wet ground. The frog didn't move. Sean lay down next to it. "Go on," he coaxed. "Go on, little guy. You're free now." The frog remained motionless. It looked calm and its color matched the Mississippi, but it wouldn't move. "Go on, little buddy," Sean said, and he picked it up and placed it on the back of his hand. "Jump now. Jump," he said, and when he lowered his hand into the water, the frog leaped off.

We went back to the table and ate our lunch without saying much more about it. When we were finished, Sean took out his tools and adjusted my gears. By the time he declared them tolerable, it was too late to ride to Hamburg and back, so we rode to the landing where the boat for Brussels boarded, pedaled up to the sign that said FERRY, touched it as we always did, and turned around.

"Ferry," Sean said as he placed his palm against the metal. I rode up beside him and balanced myself against the sign, and we sat there for a moment watching the ferry as it made its way slowly to the other side. It was the last ride we took together. Eleven weeks later he was dead.

I was still standing at the kitchen window with my mother. Other birds had come to eat the seeds she had spread for the cardinal, plain birds, brown ones and black. Four or five of them were flying to and from the windowsill, and she greeted each of them as she had the cardinal. "Hello, little guy. Hello, pretty fellow." Watching her, I wondered why I felt an ambivalence, bordering on disbelief, about some things — angels — while others, things that could be seen as equally implausible, I accepted without a second thought. A few weeks earlier, my mother had told me about the vision she had had at Mass the Sunday before Sean died. All the boys from Sean's grade school basketball team had appeared at the altar, wearing their good suits. It looked as if they might be going to a sports banquet or their grade school graduation, my mother said, but instead they were standing in two lines, carrying a casket. The vision had come to her quickly, after communion, and though she didn't have time to notice where Sean was standing, she sensed, for sure, that he was there.

I found my mother's belief that Sean was an angel unsettling, but I had accepted her vision as if it were the most ordinary of occurrences. Nor did I find it odd that she didn't recall having the vision until the day following Sean's death, that it returned to her after the fact, wrapped in its own revelation. My mother would see it as God's way of preparing her in advance, placing Sean's death somewhere in her subconscious, telling her the time was coming.

"Move away from the window," Kelly said, coming back into the kitchen from the family room, where she'd been watching TV with my father. "Step back slowly from the birds and move away from the window," she said, imitating a voice from one of the police dramas she and my father favored. "I mean it," she told us, switching back to her own voice. It was a phrase she tacked on to almost everything she said. She had been born with a forceful personality, my mother maintained, and from the time she began to speak, she seemed to possess the speech patterns to support it.

She took a can of cookies from the cupboard and headed back to the family room. "Prettiest little girl in the world," I heard my father say. Unlike the rest of us, she had never been blond, and with her dark hair and bright blue eyes — and because she was the baby — she had long ago declared herself to be my father's favorite.

Driving home from my parents' house that day, I wondered if there was any connection between my mother's vision and her belief that Sean was an angel, and I thought that if angels did exist, maybe sudden death — suicides, accidents — was the means by which they were recalled from the world, their ascensions masked in human misfortune. Perhaps this was how the movement of angels, their very existence, was kept a mystery. And if this was true, did doubt necessitate such tragic endings? Was there a time when angels came and went more freely? Maybe there was a time when angels disappeared by putting fern spores in their pockets, a simpler time, a time when people accepted — even believed in — the inexplicable, a time when everyone in the world was more like my mother.

Within a few years, I would have a vision of my own, but it would not make me less ambivalent about angels. Sick with a strep infection, I had a fever so high I was to be hospitalized the next day if it didn't go down. During the night, my fever climbing, I saw Sean sitting on the floor at the end of a long tunnel of white light. A child of four or five, he was wearing the white suit he had worn as the ring bearer at a cousin's wedding — white shirt, white jacket, white shorts, socks, and shoes — and he was drawing something with white chalk on the white ground. I was standing at the other end of the tunnel, outside the light, holding a picture. "Come closer," he told me. "I can't see it." I stepped into the light, and as I walked toward him, I could feel my body beginning to disappear. "Come closer," he kept saying. "I can't see it." Little by little, with each step I took, my body left me. It was as if I were being erased, rubbed out in clean horizontal bands beginning with my feet, my ankles, my shins, my calves. I continued to walk through the light. As each level of my body disappeared — my knees now, my thighs and hips — I felt more and more euphoric. I kept walking, wanting to be near him, to be with him again, until, just one step away, only a small sliver of me remained. I existed only above my forehead. With one more step, the step it would take me to reach him, I would leave my body completely. When I realized this, I became frightened and stopped. He held out his hand.

"I'm here," he said. "You can come now if you want."

"I can't. I'm scared," I whispered, and as I did, he disappeared.

By morning, my fever had fallen, and I lay there wondering what had happened during the night. Was it a dream? A hallucination? Would I regret not having gone? Lying there, I felt a deep sadness, as if I had lost Sean a second time, and I remembered being at my parents' house the day after he died, feeling as if every time the door opened he would walk through it. After a while, I couldn't bear it any longer. I went out in the back yard and stood in the snow, everything so white around me — the house, the ground, the trees, the fence. I stood there, coatless, watching my breath leave my body, the cold exhalation of it, thinking in the silence *Sean,* his name my only thought now, breathing out and breathing in: *Sean.* The snow settled like a wet white dust all over my body. Buried deep within it, my feet were no longer visible. I stood there, maybe hours, maybe minutes, until Mary appeared at the back door. "Come in," she said. "It's cold." And as soon as she said it, I could feel my hands and feet stinging from the very thing she called it: cold, cold.

Waking up from the white light, I realized that the light and the snow were like two sides of the same story, my body leaving me in whiteness, my body coming back. But which was more real? The dream or the day after, life or what comes later? "Come closer," Sean said. "I can't see it." "Come in," Mary told me. "It's cold." I looked at her standing in the doorway, the house dark behind her, the brightness of the snow between us, a sight so familiar — Mary calling me to come in. "Come in," she would call out when we were children. Maybe it was dark or time for dinner. Maybe our favorite TV show was starting. There had always been something maternal about Mary. She had a way of making sure we were all where we were supposed to be, Michael and I when the three of us were growing up, Sean and Kelly later. "Come in, it's cold," she called out, and looking at her in the doorway that day, I suddenly saw her as the child she had been, storing extra pairs of mittens in the mailbox when it snowed. By keeping them outside, she reasoned, covered and close by, we could exchange our wet ones for a dry pair and keep on playing, uninterrupted by the wait that came with standing at the door and calling for my mother. Walking toward her through the snow that day, it was as if the past were taking over, surrounding me, protecting me, momentarily shutting down the sting of Sean's death, the cold numbing bite of it, letting me walk toward a memory of Mary — the mailbox, the delight of dry mittens — until I entered the house again and could see the pain on my parents' faces, on Michael's face and Mary's, on Kelly's, seeing on their faces the pain that must have been on mine.

The snow had started falling the night before. Michael had driven through it, to my apartment in South St. Louis, to tell me the news and bring me back to my parents' house, where my mother, my father, and Mary were waiting. Spending the night at a girlfriend's, Kelly would also have to be picked up, brought home, and told the news, but my parents would wait until morning to do it, my mother sending me to get her, as she had sent Michael for me.

It was three or four when Michael knocked at my door. Normally a forty-minute trip, the drive had taken him hours. On the way home, he leaned forward over the steering wheel, pushing his face as close as he could to the windshield, trying his best to see through the snow. "Shit," he said as we crept along. "Snow motion," Sean called it once when we were driving together in the other direction, he, Mary, Kelly, and I slowly making our way to my apartment. "Get it?" he said, and gaining momentum from his pun, he launched into one of the games he liked to play. "What's the opposite of snow?" he asked.

"Don't start," Kelly said. "I can't stand that game."

"Neither can I," Mary said. Talking with her eyes closed and her head bent back on the passenger's seat beside me, she reminded me of my mother.

"What's the opposite of snow?" Sean mouthed to me in the rearview mirror.

I shrugged and he mouthed the answer, but I couldn't make it out.

"Stop it, you guys! I mean it!" Kelly said.

She hated Opposites, a game Sean and I had invented one day when we were riding our bikes. In it, words could be paired based on meaning, sound, or both. What is the opposite of *wood*? Wouldn't. Of *fast*? Feast. Of *hear*? Say. Of *boy*? Sink. "Bzzz," Sean would say, imitating the wrong-answer sound on a game show. "We'll have to consult the judges on that one." And then we would argue over whether the correct pronunciation of *buoy* was *boy* or *boo-ey*.

What *is* the opposite of snow? If he finally told me, I don't remember. "What's the opposite of firefly?" he had once called me at work to ask. "I give up." "Waterfall," he said, and then hung up right away, as if he realized, as I had, that *firefly* and *waterfall* were as good as the game would ever get.

It was almost morning now. Michael was still hunched over the steering wheel, maneuvering us through the snow. By the time we got home it would be daylight, and it would seem as if it had come upon us suddenly, as if we had driven out of all that white darkness directly into dawn. I watched Michael peering through the windshield, ready for any glimpse he could get of the road. We had hardly spoken, and it felt to me — did it feel that way to Michael? — that words would mean nothing now, as if, after delivering the message that he had driven through the darkness to deliver, knocking on my door to say that Sean was dead, all language, any language — Sean's language, opposites, puns, rhymes, riddles — had become meaningless.

So we were silent, and in the silence he kept steering us toward where we needed to be, just as he had years earlier with Sean, getting up in the middle of the night to head him toward the bathroom as he wandered in his sleep from room to room. Sean was six or seven then, Michael nineteen or twenty. Sharing a room, they slept in the same double bed, and when Sean began to sleepwalk, it quickly became clear that the bathroom was his desired destination. After roaming the house one night, he went back to his room and relieved himself in a drawer full of Michael's sweaters. The next night and every night thereafter, up late as I always was, it became a common sight to see them, Michael's hands on Sean's shoulders, steering him toward the toilet, aiming his penis for him while he peed, sound asleep, then guiding him back to their bedroom, the two of them never talking.

When we finally got home, my father was waiting at the front door. He held out his arms to me. "I'm sorry," he said.

"It's not your fault," I told him.

He turned his face from me, looking hurt and quietly offended.

"I'm sorry," he said, taking me in his arms and trying again, "that you're all so young." And he hugged me then, harder than he ever had.

Late in the afternoon, the snow stopped falling, and we all went together to the funeral home and the florist.

"Do you think they'll have enough flowers for him here?" my father asked as Mary and my mother placed the order.

"It's just the snow, Dad," Michael said, knowing that my father, a former salesman, took the absence of other customers as a sign that the florist was failing.

Looking less worried, he took Kelly's hand and held it.

"What's your favorite flower?" he asked her, and she shrugged, undecided for once, saying she wasn't sure.

Back home, the flowers ordered, Sean's funeral arranged, I went up to my old room and lay on the bed. The house was quiet now. No one was crying, though that would come again, my parents rushing to one or the other of us, one or the other of us rushing to them. But for now there was nothing. Next to my bed was a picture I had taken of Sean, and I picked it up from the nightstand and stared at it — Sean at twelve, wearing a T-shirt that said TRIUMPH. He was sitting in my room, on my bed, saying something, though he had stopped and smiled when I snapped the shutter. What was he telling me that day? "Did you know . . . ? Did you ever wonder . . . ?" "Did you know that on a cellular level people are basically the same?" "Did you ever wonder, since the orbit of every planet in our solar system is an ellipse, why elliptical orbits are called eccentric? I mean, geometrically, that's what they are, eccentric, noncircular, but in a larger sense it's ironic, isn't it?" "If you had to be a rock, would you want to be igneous, sedimentary, or metamorphic?" he wondered. "You'd probably be metamorphic," he said before I'd made up my mind.

"Do you think my teeth look okay?" he had asked me a few months earlier, when, after six years of orthodontic work, just before he turned fifteen, he was finally free of his braces. I was the reason he had been forced to wear them, having knocked most of his front teeth out one night when he was nine. Not realizing he was standing behind me, I had hit him hard in the mouth with a bat. It was Mary who had thought to pick up the teeth so they could be reimplanted. "They're his permanent teeth," she kept saying. Michael went to find my parents, and Mary held a wet rag to Sean's mouth while Kelly and I and the other kids who were playing baseball in the neighborhood that night scoured the street for his teeth.

He hadn't held what had happened against me. I looked at his picture. His braces had been temporarily removed, and though I couldn't remember the reason he had been spared them that summer, I was grateful that he had been given a few months' reprieve. He was smiling more than he might have.

Outside, the day was turning dark, and I got up and stood at the window. It was snowing lightly again, and the tracks I had made earlier looked like faint shadows now, sinkholes filled with fresh snow. Any sign that I had been out there would soon be covered completely. Looking down at it all — the subtle changes in the snow, my footprints disappearing — reminded me of the first time Kelly encountered impermanence, the phenomenon of things passing. She and Sean had spent the morning making snow angels. She was three or four, Sean six or seven. From my window, I watched the two of them falling backward, flapping their arms and legs, standing up to admire their creations, then falling down again. Angels? Sure, if that's what you see, I remember thinking when Michael and Mary taught me how to make them. To me, they looked like blurry abstractions, semicircular swaths in the snow. But seen from the second story, their shapes were more discernible.

I watched Sean and Kelly cover the yard, making one angel after another, until, every inch imprinted, they lay there in the last figures they had formed, each of them embodying the outline of an angel. By afternoon, Kelly up from her nap, the angels were barely visible. Buried beneath a fresh snowfall, only the slightest, most delicate indentations — half of them the size of Sean's body, half the size of Kelly's — could be seen as evidence that they existed. Rushing to the window to see them when she woke, Kelly started crying and wouldn't stop until Sean took her out to do them over, and falling down and getting up again, falling down and getting up, they resurrected all the angels they had made that morning.

Now the snow was too deep and wet for anyone to lie in, much less move. Turning dark, the yard — the world as far as I could see it from my window — was empty of angels. I could hear my father downstairs, talking softly to someone. Would there be enough flowers for him? he had asked earlier. No. Of course not. How could there be? "Sunflowers," I wanted to say when Kelly couldn't name her favorite flower at the florist's, but I couldn't think of the word then, only the image of her and Sean and Mary sucking on the seeds in the summer and spitting the shells on the sidewalk. "Sunflowers," I remembered now, looking at the yellow rectangle of light spilling out from the kitchen, coloring a small section of snow below my window.

I followed my father's voice downstairs. In the living room, he and Michael were paging through the paper. My mother was sitting at the kitchen table making a list, and in the family room, Mary was feeding Sarah. I sat down beside her on the sofa. Across from us, Kelly was rocking fiercely in the rocker next to the stereo, as she always did, the music turned up so high we could hear strains of it seeping out of her headphones. "What?" she would say, lifting the headphones off her ears a bit if I looked at her too long. "Nothing," I would answer.

It could have been an evening like any other, with Sean coming home any minute. "Why's everyone here?" he would ask, before turning his attention to Sarah, squeezing in between Mary and me on the sofa, saying it was his spot, just as we had done fifteen years earlier, Michael, Mary, and I claiming the seat closest to Sean, arguing over whose turn it was to feed him or hold him or rock him to sleep. "There's time," my mother would always say to settle it, taking him from one of us and giving him to another.

Yes, there's time, I thought. There was the time before he was born and the time after. Ordinary time. A time when we woke up every day, our souls still within us. And now there was this time. The time being. A time for which my father had said he was sorry, one for which we were all too young. It would be a time — this time — unlike any that had passed before. A long time. A time presided over by angels perhaps, messengers in slow motion.

That's what I was thinking when my mother sent me out soon after to tell the neighbors the news. I was thinking about time — the slowness of it one day, the speed of it the next. It was nearly six o'clock. When we were children, the whole neighborhood grew quiet at that hour, everyone having been called to come in. If it was still light out, the street would come alive again later, in spring and summer and early fall, all of us gathering at the streetlight after supper, claiming as much of the day as we could. But in winter, the quiet continued. No one came out again. It was quiet now, the street calm and white and quiet. As I walked from one neighbor to the next, I remembered running down the street the night my parents told us my mother was pregnant. It was a spring night after supper; Sean would be born the following fall. The news was supposed to be a secret. "It's bad luck to tell everyone too early," my mother explained. It was a Saturday night. She and my father were getting dressed to go out and I lay on their bed pleading — "Just one person!" — until my mother consented, saying we could tell the neighbor who was coming to sit for us that night. Of all the adults in the neighborhood, Mrs. Fallon was my favorite. I ran down the street and knocked on her door. She was an old woman who lived alone and spoke in one-word exclamations. "Marvelous!" she said when I told her. ("Beautiful!" she said the first time she saw Sean.) I ran back, breathless, and reported her response to my mother. She shook her head and laughed. "I meant you could tell her when she came tonight, not that you should run down there right away." "Oh," I said, and I went out to play again, satisfied that I had told someone, but still harboring the thrill of a secret, the knowledge that when the time came, I would get to deliver the news to another neighbor, and another, until everyone knew.

If she were alive, what would Mrs. Fallon say these fifteen years later? Was there a word that would sum it all up, one that I could race back and report to my mother? Or would I leave her as I left the others, stunned, speechless, all of them having peered through their front windows when I rang the bell, all of them having turned on their porch lights. Recognizing me, they opened their doors wide. But I was no one they knew that night. Standing before them, I was the darkest, most diminished of souls, a messenger of agony, coming to announce the end of the world as I knew it and the beginning of eternity, a time unlike any I had ever borne.

I stood at the end of our street, each neighbor notified. A winter night, quiet, so quiet. In another season — spring, summer, fall — we would come out again and gather at the streetlight, children telling stories, choosing sides, my team, your team, each of us running to reach the place called home — the streetlight, where every game began and ended — yelling *Safe!* when we got there, the word ringing out over and over, one after the other of us touching the streetlight, safe, safe, safe.

But it was winter now. The quiet would continue. No one would come out again. I stood at the end of our street, snow falling all around me, and I watched in the darkness as doors started to open, neighbors coming out, one and then another, two or three of them walking slowly up the hill to our house.

From a distance, I followed them back home, pausing to watch the snowflakes swirl beneath the streetlight, white glitter falling to the ground. "Did you ever wonder what life would be like if there were only two states of matter instead of three?" The snow looked like white sparkling dust circling the lamppost. Wasn't that how stars got started — cosmic dust spinning in a circle? No two snowflakes are the same, we were told as children, but I didn't believe it. How could anyone really know? How could we be certain there wasn't another one out there, identical, waiting to drop? Over time, how could we tell? I watched the snow falling through the light. It looked beautiful and endless. What is the opposite of eternity, I wondered. A short life, he might tell me. A story still beginning.

Two Summers

I began the summer of 1966 — the summer before Sean was born — by hiding behind a refrigerator in a store that sold household appliances. It was nearly nine o'clock in the evening and I had already been timing my parents for ten minutes to see how long it would take them to notice I wasn't there. They were pricing washing machines. My mother told the salesman she was certain our old one wouldn't last long doing diapers every day, and when he started asking about the size of our family and what kind of habits we had — getting a sense of our lifestyle, he called it — I decided to disappear.

I had never been behind a refrigerator, but it felt a lot like hiding under a bed, only standing up. About the same amount of space that lies between a bed and the floor stood between the refrigerator and the wall, and the electrical coils that snaked from top to bottom along the back caused the same stifling feeling I got from bedsprings when they were above me. It made me halfway hold my breath.

I had on my white uniform shirt from school and pink stretch shorts that my mother said were a little too snug, which is why I remember it was the very beginning of summer — the first night of summer, in fact. Ordinarily, we weren't allowed to wear our uniform shirts after school. My mother said it wasn't a good habit to get into. I took this to mean she didn't want us to look like the poorer kids in the parish, who never changed out of their school clothes and sometimes even wore their white shirts on weekends.

But in my case, I knew it was more than that. I was hard on my clothes. This was one of the first facts I learned about myself. I overheard my mother telling it to one of the nuns the day we donated our clothes to the clothing drive. I wasn't in school yet — I must have been five — so I accompanied my mother wherever she went. It was a fall day and we drove up the hill to North American Martyrs, the school Michael and Mary already attended, with grocery bags full of clothes in the back seat. I was particularly proud of myself that day because I had figured out, without asking, that a clothing drive was called a clothing drive because you drove your clothes to it.

I helped my mother carry the clothes into the cafeteria, where a nun named Sister Agnes Charles inspected each piece and sorted everything into separate piles. I thought at first she was sorting the clothes according to the countries to which they would later be sent. In my mind, all of my father's shirts were going to Southeast Asia. Michael had told me all about the missionary countries where the men worked shirtless in the hot sun and the priests had trouble converting them because they didn't have any clothes to wear to church. We even played Rice Paddy one day. Michael stretched the hose all the way to the end of the back yard and let the water run until a small square of grass was flooded. I took off my shirt and walked barefoot through the rice paddy, poking at the ground with a stick, while Michael paced back and forth beside me, slightly bent, saying some prayers. He knew a little Latin. We had to quit, though, before I was converted, when the rice paddy started to seep under the fence into the Pollos' yard next door, setting their doghouse afloat.

So I was sure the pile Sister Agnes Charles made of my father's shirts was headed for Southeast Asia. The pile she made of my mother's evening dresses I wasn't as certain about. We had carried in a lot of bags, and Sister Agnes Charles was creating more and more piles around herself.

"They're all clean," my mother said as Sister Agnes Charles put her hands in the pockets of each pair of pants. "And none of them needs mending."

"We'll see," Sister said as she scrutinized the seams on my mother's old skirts.

My mother and I stood hand in hand before her: Sister Agnes Charles was on one side of the table; we were on the other. Her hands moved quickly over our offerings, as if it were a sin to touch what we had placed before her, each piece of clothing tainted by the relationship it bore to our bodies. I watched her, dressed head to toe in her black habit, and wondered how long it had been since she'd worn real clothes. The next year, she would become my first-grade teacher, and on the rare occasion when she stopped behind my desk and put her hand on my shoulder, I would squirm, remembering the clothing drive and the judgments she had passed upon us.

Soon I realized that the piles Sister Agnes Charles was creating had nothing to do with geographic destination. She was sorting by sex and size, and since this interested me less than imagining how people in foreign countries would look wearing my parents' clothes, I let go of my mother's hand and went over to the drinking fountain to see whether the water in it tasted different from the water in a department store. Michael said the water that ran through the pipes at North American Martyrs was blessed. When I finally turned off the fountain, I heard my mother say, "She's hard on her clothes." She was explaining why we didn't have more girls' stuff to give.

I felt shocked by this, and ashamed. My mother had never told me I was wearing out my clothes too quickly. But at the same time, I knew immediately that what she said was true. Her words fit me perfectly, like an aspect of myself I had been waiting for someone to name. And I remember feeling different, as if suddenly I had changed into someone who had an identity, someone whose qualities could be described. As a child, as an adolescent, and even as an adult, I would never be fully aware of physical changes within me — my body developed without much notice on my part — but the kind of transformation I experienced that day, when I was introduced to a new definition of myself, was one I would be aware of again and again. In that same week, I would learn that I was lucky, too quiet sometimes, and tall for my age, but after what seemed like an initial awakening, the revelations slowed down, leaving long periods when nothing about me seemed new or noticeable.

In the car on the way home, I wanted to ask my mother if I had been born hard on my clothes, but just before we left the cafeteria, Sister Agnes Charles pointed to the pile of my mother's old evening dresses and said she didn't think they were appropriate. My mother nodded, not in agreement, but as if she were acknowledging Sister Agnes Charles's authority in the situation, and she picked up the pile of dresses, took my hand, and led us out.

In the parking lot, before starting the car, my mother looked at the dresses sitting on the seat between us. The one on top was a color my mother called aqua, and the neckline was trimmed with silver sequins and pearls. She ran her fingernail over the line of pearls, and when one came off she started crying. "I just thought these might lift someone's spirits," she said. She put the pearl in my hand and bent her head to the steering wheel, hitting the horn. The sound of it brought Sister Agnes Charles to the window.

The sun was strong that day, and because of the brightness outside and the fact that the cafeteria windows were tinted, everything inside the school seemed coated in darkness. Even when I squinted, I couldn't make out Sister Agnes Charles's body. Clad in black cloth, she receded into the room around her, and only her face, lit by the white trim of her headdress, shone through. I watched her face floating in the window, pinched the pearl between my fingers, and waited for my mother, still crying beside me, to sit up and start the car.

All of this is to say I'm sure it was the very first night of summer when I hid behind the refrigerator three years later. By the age of eight, I hadn't grown any easier on my clothes, so it was only on the last day of school that my mother allowed me to stay in my uniform shirt past supper. I don't remember, really, why I thought it was such an accomplishment to keep it on.

Behind the refrigerator, I leaned against the wall and looked out the window while I waited for my parents to wonder where I was. Outside, the lights on the parking lot started their ascent, from dim to bright to brighter, as if they themselves were responsible for pulling down the darkness. At the end of the parking lot, a strip of blue neon blinked OPEN TONITE 'TIL NINE, and across the highway, a traveling carnival was set up at the shopping center. I watched the illuminated spokes of the Ferris wheel spill forward, starting, then stopping, and imagined the seats swinging back and forth as people were let off and let on, until, finally, the wheel went around once uninterrupted, and then a second time, and a third, carving a slow circle of white and yellow light into the summer sky.

As I watched, my parents walked past, and for the first time I noticed that my mother was pregnant. She was standing with her side to me, the window was between us, and the place where her body bulged out filled the bottom of the first o in a line of letters that spelled NO MONEY DOWN across the width of the window. My parents were outside. They weren't looking for me. They were leaving without me.

I knocked on the window — maybe they were bluffing, I thought — and waited for them to turn around and laugh. Instead, they stepped off the sidewalk and headed toward the car. Just then, someone began shutting off the lights, while the salesman who had waited on my parents wheeled in a sign from the sidewalk and locked the front door. As I bolted from behind the refrigerator, my shirttail got caught in a coil, and the salesman looked up, surprised. Instead of coming to help me, he stood at the door, scowling, and when the last set of lights went out, he shouted, "Hurry up!" My hands were shaking. I couldn't free myself. I banged on the window, hitting my hand against the word where my mother had been, but my parents went on walking. In a panic, I yanked my shirt from the refrigerator, leaving a large piece of it stuck in the coil, and ran past the salesman and out the door.

I hurried toward my parents, tucking in my shirt so they wouldn't see that it was torn, and when I caught up with them, I took my mother's hand. They didn't seem surprised to see me. They said nothing, in fact, and my mother's hand registered no response when I took it, aside from allowing me to hold it, as I almost always did when I was with her. And each time I think of that night, I remember how cold it had become. The temperature had dropped quickly, as it often did on those evenings in early summer, and as we walked toward the car, I looked at the Ferris wheel piercing a hole in the sky ahead of us, and I pictured the people riding it — they were probably sitting close together to keep warm — and instead of taking my place in the back seat when my father opened the car door, I climbed in front ahead of my mother and sat between my parents for the short ride home.

A few weeks later, when the new washing machine arrived, Michael, Mary, and I watched the delivery men bring our old machine up from the basement and wheel it, like a large invalid, through the living room, out the door, and down the driveway. Then the men repeated the procedure in reverse, wheeling the new machine, still in its shipping box, up the driveway and into the house. We had never had an appliance delivered to our house, and it surprised me to see it arrive in a box. I had expected the machine to come off the truck hulking and exposed, a gleaming white magnificence. Shrouded, it looked instead as if the men were smuggling something in, something that could not be unveiled a second too soon. Still, I joined Michael and Mary in the self-importance that the situation called for and rattled off the machine's features to all the neighborhood kids, holding out my arms like a crossing guard to keep them from getting too close.

In the basement, after the men had finished hooking up the hoses, the bigger man bumped against the table that held a cage where we kept a white mouse. The cage fell to the floor and the mouse darted out. Seeing it, my mother screamed, not because she was afraid of the mouse but because the men had neglected to replace the cover on the sewer hole. I was standing next to the hole — in diameter, it was no bigger than a baseball — and my mother kept screaming and motioning to me to put my foot over it. But I couldn't figure out what she was trying to tell me. It was as if we were all frozen in place — the men, Michael, Mary, my mother, and I — and could do nothing but witness the white mouse streaking across the floor toward the sewer hole. When it reached the opening and the ground suddenly disappeared beneath its body, the mouse seemed to splay out, as if it were about to glide over the hole, but then it dropped down into the darkness like an inverted parachute, its stomach sucked under, its pink feet sticking up in four perfect points.

Horrified, my mother put her hands over her eyes and started to cry. Later in our lives, Michael and I would see this same kind of reaction in our mother when we discovered one summer that we could cause slugs to evaporate by covering them with salt. Michael said they evaporated because they were basically just live sacks of slime and dirty water. My mother found us kneeling over some on the sidewalk with a container of Morton's salt. We thought of it as a science experiment, and when Michael told her to watch, she screamed and covered her face with her hands again. Crying, she grabbed the salt from Michael and pulled us up to our feet by one arm, telling us to get the hose so we could spray the salt off the slugs before they lost their lives completely. Afterward, she said what we did was like pouring lye on humans, and though that analogy made us ashamed of ourselves, it didn't sink in. Later, we would admit to each other that we did not share our mother's sympathy for slugs, and when we grew up and Michael had a house of his own, sometimes in the summer, when the slugs were particularly pervasive, we would think nothing of passing an evening together sitting on his porch swing, talking, and getting up to mindlessly salt them one or two at a time.

There would be similar incidents throughout our lives, times when our mother caught us pulling the light out of lightning bugs or holding a finger down on one leg of a daddy longlegs, watching it struggle to get away. But that day when the white mouse fell down the drain, we realized for the first time the extent of our mother's affinity for small creatures. Mary tried to console her by explaining that the mouse certainly didn't die, that it would be just as happy, probably more at home, living in the sewer line. I put the cage back on the table and swept the cedar shavings into the sewer hole. My mother said that was sweet, and I think she meant it was good of me to send the shavings in after the mouse, but that idea hadn't occurred to me. It was just a convenient way to clean up. Michael tried to change the subject by asking her if she wanted to save the big washing machine box for the baby. It was standing next to the dryer like a discarded cocoon. I was confused by this. At first I wondered what baby he was talking about, and then I realized he was talking about our baby, as if it had already been born. Until that day, I hadn't thought of the baby as actually being one of us, someone we should be saving things for, and beyond that, I didn't understand what a baby would need or want with a big box. Michael told me later that it would be a good box for the baby to crawl in and out of, like a cave. "They like openings," he said. My mother said it would be a long time before the baby would show any interest in the box, and she suggested we do whatever we wanted with it, she didn't care, she was going to lie down.

And so I knew there was nothing we could do to make her feel better about the mouse, though I wished with all my heart there was, because when my mother lay down in the middle of the morning, after we were already dressed, it was the sadness sinking down in her, the sadness that started when her father died. When I was younger, in the years before I went to school, she would take me back to bed with her after everyone else had left, and we would lie close together, with only our breathing between us, and I knew always to be quiet, because she was not sleeping, she was saying her prayers. And that is the first thing I learned about love, lying as close as I could to my mother, knowing that in some way I was her safety, that in some way I was a good reason for her to finally get up. So on days like the one when we lost the mouse, it became my habit to slip into her bed to see whether she was sleeping, and when she wasn't, she would ask me with her eyes closed what time it was getting to be, and I would tell her, and almost exactly fifteen minutes later she would get up.

Michael and Mary followed my mother upstairs and then went outside to do whatever they did on days like that. I sat under the steps in the basement for a while, sorting through a box of seashells my grandmother had given me, and then I pulled the washing machine box into the back yard, put it under the crabapple tree, crawled in, and fell asleep. When I woke up, I heard my mother calling us all to come quick. Michael and Mary beat me to the bathroom, where my mother was sitting on the side of the tub. Inside the tub, a mouse ran round and round as if it were grinding a white line into the green enamel. Mary said she felt she was being hypnotized. My mother put her hand down in the tub, and when the mouse ran over it, it slowed down a bit, and then a bit more and a bit more, until it finally broke out of its trance and stood still. My mother picked it up and declared it reborn, and since we had owned the mouse only a few days and had never agreed on what to name it, my mother christened it Maytag and carried it with her when she went downstairs to do a load of wash.

That night Mary and I slept outside in a structure we built using the washing machine box and several blankets. Mary called it a dwelling. It fanned out from the far corner of the back fence, which bordered a row of honeysuckle bushes on one side and a big snowball bush on the other, making it a masterpiece of privacy.

Mary brought out her transistor radio and we moved the plastic dial from station to station, trying never to land on a song we didn't like. We were lying with our upper bodies mostly in the box. Above our heads, we cut a flap in the cardboard that we could open and close with a stick, depending on whether we wanted to see the stars or sleep in total darkness. The moon was directly above us, so low in the sky that it looked tethered to our cardboard box. At our feet, the blossoms of the snowball bush formed a soft white wall, and above our heads, honeysuckle hung heavy and sweet.

Before we got tired and stopped talking, I asked Mary why babies liked openings. She said she didn't think she should tell me. I asked her whether she wanted our baby to be a boy or a girl, and she said she just wanted it to be healthy. Mary was ten that summer and already had a habit of sounding grown-up.

"Why wouldn't it be healthy?" I asked. I didn't understand how a baby could be born any other way. I thought it took being in the world a while for someone to get sick.

"All sorts of things could happen," she said.

"Like what?"

"Things," she said. "I guess if you want to know, I'd rather have a boy. But I'm not making any bets with you and Michael."

"Why?"

"Because betting is vulgar," she said. *Vulgar* was a word Mary said a lot that summer, in sentences and just by itself.

"I mean, why do you want a boy?"

"Because I already have a little sister."

She said this without sentimentality, as if she were describing her status to a stranger or citing a numerical fact, and it took me a minute to realize she was talking about me. I always thought of myself as Michael's little sister, but it never occurred to me that I was Mary's too. I guess this was because we were the same sex, and with only two years between us, I wasn't nearly as young next to her as I was next to Michael, who would turn thirteen when the summer was over. Besides, I always thought of us as an indistinguishable unit. We were the girls in the family, which to me made us equal.

And so for the first time, I thought of the unique position I was in. As I figured it, since I had neither a little sister nor a little brother, I was the only person in the family who had no reason to want our baby to be born a certain sex. For Michael and Mary, a boy would be something other than what they already had. And for my mother and father, a boy would probably be better, since with two daughters, two sons would add some symmetry. I'm sure that wasn't important to my parents, but that summer, as I slept outside with my sister, it made a certain kind of sense. So for the first time since my mother had announced to us that she was pregnant, I began thinking of the relationship I would have to the baby and the relationship it would have to me, and the fact that I had no preference for a boy or a girl felt like a gift I could give, a gift that was my own and no one else's: the ability to want someone who was no particular way whatsoever. But at the same time, I knew deep down that my not caring whether the baby would be a boy or a girl extended far beyond its birth. Having no preference, not being swayed by one sex or the other, was a basic part of me that at eight I had already felt but could not begin to name. And I knew this somehow made me different. Michael and Mary had a certainty about themselves that was missing in me.

Mary asked me what did I want, a boy or a girl?

"I don't know," I said.

"You want a boy too," she told me.

"Maybe," I said, and we lay there quietly for a while, watching the moon move over.

"Do you want to play Beach Blanket Bingo?" I finally said, thinking it would be a nice way to end the night.

"I guess," Mary said.

"Can I be Annette Funicello this time instead of Frankie Avalon?"

"No," she said, "you can never be Annette."

"Why not?"

"Because I don't want to be the boy. I have to be the girl."

"Please," I said. "I know how to do the girl part."

"No, you're better being the boy."

"Then I'm not going to be Frankie Avalon," I said. "I'll be some other boy. The skinny, pimply one with glasses who always reads books on the beach."

"Oh, all right then. Be Annette," she said. "But this is the only time you get to switch."

Mary pulled the flap in the top of the box closed and we snuggled together and said things as though we were on a date. Then Mary lay on top of me and we kissed wildly with our mouths closed and rolled around a bit in the box. In Beach Blanket Bingo, playing the girl was no different from playing the boy, but that night, for some reason, I just had to be the one to cry out, "Oh Frankie, Frankie!"

That summer was the first summer my mother didn't take us swimming. We still went swimming on our own, but it wasn't the same as going with my mother. She was an accomplished swimmer and diver, and her confidence and ease in the water rippled over us, making us carefree and buoyant. When we went swimming, she never stopped smiling, and she splashed and laughed and enjoyed us completely. Going to the public pool with her was pure happiness, and watching her dive off the high dive, with her legs straight and her feet pointed, opened up a place in me that no other feeling could match. At the end of those afternoons at the pool, when my mother came out of the water, pulled off her swim cap, and shook out her hair, I couldn't imagine how anyone could possibly exist without a mother like mine.

As some sort of compensation that summer, my mother signed me up for swimming lessons. She herself had taught us all the strokes, as well as how to kick and breathe correctly. Michael and Mary were good swimmers, but I had a tendency to sink.

"I can't take swimming lessons," I confided to Mary the night before I was supposed to start.

"Why not?" she asked.

"I don't know how to swim."

The next morning, my mother said she would walk over with me. I told her there was a sign at the pool that said parents weren't allowed to watch, so it would just be a waste of her morning. I would be fine by myself. When I got there, I wasn't going to go in, but then Michael rode by on his bike and said I looked stupid just standing there. If he had ridden off right away, I would have hidden, but he wanted to hang around, he said, to see if anyone would drown.

The instructors made us line up at the side of the pool and asked questions about what we could and couldn't do, then grouped us according to when we raised our hands. The first question they asked was who could already swim the length of the pool. Several kids raised their hands to be in that group. I was one of them. We were called "advanced."

It wasn't completely true that I couldn't swim; I just didn't look too good doing it, and it took me so long to get from one point to another that the instructor usually stopped me halfway to say we were going on to the next stroke. I worked hard and tried not to notice what anyone else looked like. Even harder, I tried not to notice anyone noticing me. Early on, one of the instructors suggested that I might benefit more from the lessons if I dropped back a few groups. But he didn't insist, so I stayed and struggled against the odds that I would ever improve.

On the day of the last lesson, we had to pass a test if we wanted to get a certificate. The test for the advanced group included diving off the high dive. Climbing the ladder, I noticed Michael standing with his bike on the other side of the fence. I had jumped off the high board before, but I'd never dived. I walked to the end, looked down at the water, and went for it head first. It was the sound of the water, before the stinging sensation, that stayed with me after the first belly flop. Without asking, I stood in line to try again. Michael was still standing at the fence, and I made myself not look at him. This time I took a few steps and sprang off the board, but I couldn't straighten out fast enough, and my stomach hit the water bluntly. When I climbed up a third time, Michael was sitting on his bike with one hand on the fence. I thought about looking at him, but decided not to. I went to the end of the board and bent my knees more, but that made me flip over, and I met the water with the same force as a belly flop, but on my back. When I went up the fourth time, all the advanced swimmers were standing around with towels on their shoulders, waiting for their certificates, and Michael was gone from the fence. He went home, I learned later, to tell my mother he thought I was trying to kill myself. By the time he and my mother arrived at the pool, the instructor had said ten times was the limit and made me stop trying because bruises were beginning to turn blue all over my body.

I was standing in front of a mirror in the shower room when my mother walked in. In the shower room, I was always afraid my feet would slip out of my thongs and touch the floor. It was the worst room in the world, the most public part of the public pool, where everything was wet. I was looking closely at my legs for the first time. They were full in the front, and their shape, from my knees to where my swimming suit started, reminded me of sails billowing out on a boat. Earlier that year, an aunt had given Mary and me fishnet stockings. Mine were too small. Michael was with my mother when she returned them. When he came home, he said they'd had to go to the chubby department to find a pair my mother thought would fit. My mother assured me she had been chubby too as a child and had slimmed down when she started diving. But it didn't look as if diving would play a part in my future. When my mother joined me in the mirror, she kissed my cheek. "Will I always be this way?" I asked.

"What way?"

"The way that I am."

"I hope so," she said, and she pulled me close to her, even though I was wet, and we looked at each other for a moment in the mirror. My mother picked at the fabric of my swimming suit where she noticed a hole starting near my hip. "Remind me to mend that," she said, with her lips pressed against my head. I wanted to ask her if she was ever afraid of anything, but the baby kicked me in the back.

I turned around, startled. "What was that?"

"The baby," my mother said. She took my hand and held it on her stomach. Nothing happened, and then it happened again.

"It's kicking," she told me.

"Maybe it's punching."

"No, they move more with their feet," she said, "like kicking under water."

"What water?"

"The baby's surrounded by a sac of water," she explained. "To keep it safe."

"It's under water in there?"

"Yes," she said. "It's floating."

It kicked again, and she moved my hand a little lower on her stomach.

"If we live in water all those months, why aren't we born knowing how to swim?" I asked.

"Maybe we are," she said.

My mother believed in miracles, in the possibility of things being perfect. I had never experienced a miracle myself, but my mother had experienced many. I knew if someone threw a newborn baby into the pool, my mother would be the first to jump in and save it. But I also knew if the baby started swimming and saved itself, that would be something my mother could easily accept.

"Does it hurt to have a baby?" I asked.

"Only until you see its face," she said. "Then you forget."

"So it does hurt."

"When you're older, you'll be ready for it."

"Who hurt you most?" I asked. "Me, Michael, or Mary?"

My mother laughed, took the towel from my shoulders, and started drying my hair.

"Maybe me," I said. "I hurt you last."

"It looks like you hurt yourself." She rubbed her hand over the bruises under my arm. "Why were you doing this?"

"Doing what?"

"You're always doing this," she said, and she took my chin in her hand and made me look at her. "Stop doing this," she said, and she hung the towel around my neck and started to cry.

We had been alone in the shower room, but just after my mother began to cry, a short, heavy woman wearing a plain black bathing suit walked in. My mother was sitting on the bench by then, holding my hand, and I was standing beside her with my bruises. The woman stood under a shower and turned the water on without bothering to close the shower curtain. As she peeled off her suit, her flesh seemed to inflate. I was facing her, but my mother was sitting the other way and didn't see her. The woman hung her bathing suit on the faucet and then stood with her back to the shower. I had never seen a naked woman. She bent her head back to let the water run through her hair while she rubbed a bar of soap between her legs. When she straightened up and stopped soaping herself, she turned around several times, shifting from foot to foot, and it wasn't until she was facing forward again that she opened her eyes and saw me staring at her. She glanced down at my body. Embarrassed, I pulled my towel from my shoulders and wrapped it over my swimming suit. This movement caused my mother to look up, and she saw the woman, fully naked, in the mirror. By then, the woman had turned off the water and was squeezing back into her suit. My mother got up and steered me toward the door. When we got outside, she said the woman was immodest.

Later that day, I told Michael and Mary that the baby had kicked me in the back, but they didn't believe me. Mary said it sounded like a story I made up to explain to people why I had all those bruises, and she said, "Lying's your worst sin." I asked them if they knew what *immodest* meant, and Michael looked it up.

"Not modest," he said, shutting the dictionary.

"Give it to me," Mary said, and she looked up *modest*.

The only definition we understood was "limited in size or amount."

"So *modest* means small," Michael concluded.

"So what," Mary said. "She wants *immodest*."

"So what," Michael said. "*Not* small. No one's that stupid."

So I guessed my mother was referring to the woman's weight when she said she was immodest. And I guessed that meant my mother thought I was immodest too. And that night in bed, after Mary fell asleep, I lay awake wondering what my mother meant when she said she hoped I'd stay the way I was. Did she like it that I was chubby and had to struggle at swimming? What was it in me that made her happy? What was it that made her cry? When I was a few years older and everyone in our family had been born, Mary would tell me that our mother looked into our faces at night while we were sleeping, each of her children in turn, to see whether she could tell which one of us would become sad.

"How do you know?" I asked.

"I woke up one night when she was looking at me."

"So?"

"So, she sat on the edge of my bed and told me."

"Told you what?"

"That the sadness is a sickness that runs in our family. On both sides."

We were lying in our beds. It was a winter night close to Christmas, and through the window I could see the illuminated manger scene under the red-and-green Star of Bethlehem that our backyard neighbors had mounted on their roof. One of the Wise Men flickered on and off and then burned out altogether.

"How does she know one of us will get it?" I was twelve by then, and Kelly had just been born. My mother had five faces to look into.

"Because there's one in every family," Mary said.

"I wonder who it is."

"It's you."

"No it's not," I said.

"Yes it is. You're the third."

"What?"

"I figured it out. It's always the third. Think about it," she said.

My mother was the third in her family. An aunt that we knew of, also the third. My grandfather was one of three boys, but I wasn't sure whether he was the youngest.

"It doesn't have to happen that way," I said.

"You show all the signs."

"I do not," I said.

"It's you," she said, definitively, from her bed to mine.

"Well, if you have it all figured out," I challenged, "why don't you just tell Mom so she can get some sleep at night?"

"Because I don't want to break her heart," Mary said in a sassy way. If she were standing, she would have put her hands on her hips as she said it.

I didn't say anything, and a few seconds later Mary's pillow landed on my head.

"Are you making this up?" I said from my side of the room.

"Everything I tell you is true."

"Hardly," I said. My bed was against the wall and hers was under the window. I threw her pillow back, and it hit the windowpane before it fell with a thud on the floor.

"You better watch yourself," she said, and she picked up the pillow and fell asleep.

I lay in bed and thought about the possibilities of what Mary had told me. I didn't believe what she said about sadness always being the fate of the third child, but what she said about my showing the signs was more difficult to dismiss. My mother had told me many times that I was too sensitive. Like her, I was quick to cry, and I often felt a sadness that seemed similar to hers, one that came upon me suddenly, with the force of a great fatigue. Maybe Mary was right. Maybe it was me. I lay there with my eyes open for a long time, worried and unable to sleep. It was after midnight, but the manger scene was still burning brightly on our neighbor's roof, and though I was staring right at it, I didn't even notice when the Star of Bethlehem stopped shining.

With my swimming lessons over, the summer passed quickly. Only a few other things made it memorable. In early August, the creek across the street started to rise, and after a few days it looked like a small river. For all of our lives, the creek had never been more than an idle ribbon of water at the bottom of a small ravine. When we cut through it to get to the public pool, we rarely got our feet wet. My father said the change in the creek that summer was caused by the combination of heavy rains and some construction work across town, where another creek had been dammed to make way for a new subdivision. My father had sold the bricks for the subdivision and he told my mother the houses would be nice for young families with small children.

"Are we a young family?" I asked him as we watched the creek rage past us. My father liked to watch the weather. He told us this was because he was born during the Great Tornado of 1927, and the benefit we inherited from this fact of his birth was that we got to watch the weather with him. Instead of taking cover or going to the basement if a storm warning or weather alert was announced on the news, we usually stood on the porch to see the sky change color. None of us was ever afraid when we were with my father, because the more turbulent the conditions were, the calmer he became. This he also attributed to being born during the tornado, and he told us that the moment he was born, the twister uncharacteristically changed course, moving away from St. Louis, and he was credited with saving the city substantial sums of money in cleanup and reconstruction.

"I'd say we're an in-between family," he said.

"Did you and Mom become a family when you got married?" I asked. "Or did you only become a family when Michael was born?"

"That's a good question," he said without answering.

We were standing at the side of the creek. For the past two nights we had walked over together, across the street and behind our neighbors' back yards, to see how high the water had risen. It was raining, and we stood on the bank beneath my father's umbrella. Ahead of us, behind a group of willows, the creek bent abruptly, turning the water into rapids.

My father lit a cigarette and threw the match into the creek. The rain was soft and steady, and each time he exhaled, the smoke lingered a little under our umbrella. I was still thinking of the new houses being built for young families with small children, and I asked my father whether the baby would make our family younger.

"Just bigger," he said.

"So those houses wouldn't be nice for us?"

"What houses?"

"Those houses you told Mom would be nice for young families with small children. The ones that are making the creek rise."

"Oh, no," he said. "Those are low-income houses."

I didn't know what low-income houses were, but I was relieved that they were wrong for us. I liked the house we lived in and didn't want to move, but Michael and Mary said we probably would, after the baby was born.

"Dad, don't you think we have the best house on the block?" I asked.

"All the houses on the block are the same," he answered.

"That's not true," I said, touching his leg to stop him. We were walking toward the willows and he had one arm around me, holding me close so we'd both stay dry under the umbrella. "They're all different."

I stepped back to face him, and the rain made me squint. "They're all really different," I said again.

"Some are kept up better, sure," he said, "but other than that they're the same."

We were the only people outside. Behind my father, the backs of our neighbors' houses were all cast in the color of early evening and people appeared and reappeared in their windows, walking from room to room. I wanted to point out the details that made each house different, but they suddenly seemed insignificant. In the rain, everything became equal, and from one house to the next, the windows framed scenes so similar — people talking on the phone, eating dinner, watching TV — that the houses looked as if they were connected. They weren't just the same style; they were the same story: simple, ordinary, and sufficient. And though I didn't totally agree with my father — he viewed houses as the end products of construction, seeing beyond everything extra to the basic way they were built — I realized, standing there with the rain all around us, that it wasn't so much our house that I liked as all the things that happened when we were in it.

My father threw the end of his cigarette into the creek, and we watched it succumb to the current. He pulled me back under the umbrella, as if he understood everything I was thinking, and we walked on toward the willows. We had taken only a few steps when Michael ran out from around the bend, holding a big fishing net. He wasn't on the bank; he was running above the water on a small, irregular ledge that jutted out from the side of the ravine. Running as if he were being chased, he turned sideways every few steps, looked over his shoulder, then ran a little farther. Before the ledge ended, he stopped and held his net out over the creek. Just then, an entire cavalcade of toys rounded the bend. Within seconds, a tricycle swept by us, bobbing up and down in the water, followed by a hula hoop, roller skates, a flood of dolls of all ages, going up and under as if they were gasping for air. Toys carpeted the whole width of the creek, so many, and moving so fast, we couldn't begin to count them. Croquet mallets and plastic golf clubs. A pink two-wheeler with training wheels. A baseball bat. Trucks, guns, baby toys. Seeing all this, we forgot for a moment about Michael, until a Fisher-Price cash register flew through the air and landed a few feet away from us, followed by a clear plastic ball with a family of ducks floating inside, and then a teddy bear, a drum, a sand bucket, a boat — each landing on the bank, one after another, while in the creek below toys kept coming, inexplicably, around the bend.

My father threw down the umbrella and ran along the bank toward Michael, dodging the toys he was flinging from his net. I ran for the umbrella, but it cartwheeled away from me, joining the odd flotilla and sailing downstream. My father yelled at Michael, and he looked up, surprised, and almost lost his footing. My father yelled again, and Michael lifted the net, defeated, and mounted the mud wall, grabbing my father's hand to gain the top of the bank. When he was on solid ground, my father took him by the shoulders and shook him, slapping him once on the side of the head. Michael explained that he'd been under the willows catching tadpoles. When he saw the toys coming down the creek, he ran to a clear place and started fishing them out for the baby. When my father heard this, he began kicking everything back into the water, yelling at both of us that we didn't need to provide things for the baby, that the baby would get everything it needed without us. Because my father included me in his rage, I felt guilty. He picked up the cash register with one hand, waving it in front of our faces, and asked Michael if he had considered the possibility that he could have drowned going after such crap, and then he threw the cash register into the creek and walked away.

Michael had his head down, and I could see through the rain that he was crying. Behind us, the rush of toys was tapering off. I watched my father jump over a neighbor's fence and disappear between the houses. Michael and I stood there for a moment, and then, without looking back at the creek, we walked home together and stripped off our wet clothes in the kitchen. And though we never spoke again about it, I often wondered whether Michael ever thought, as I did that day, that things might have turned out differently were it not for the fact that, unlike my mother, my father could not swim.

The next night my father took us to the movies while my mother went out with her friends. We had learned by then that the toys in the creek belonged to five families whose back yards had been washed out upstream. My mother said that should teach us a lesson about leaving our toys lying out. I wasn't sure how that lesson pertained to us, since our yard didn't back up to the creek, but my mother was putting on her lipstick when she said it, and the pleasure of watching her prolong that process was worth enduring whatever point she wanted to make.

"What movie are you going to see?" she asked when she finished coloring her lips.

"Kiss me now," I said.

"Who's in that?"

"No. Kiss me now, before you blot your lips."

I was sitting on the side of the tub, and she bent toward me to kiss my cheek.

"No. On my knee," I said, "so I can see it."

"How about your hand?" she suggested, taking it in hers and kissing it. "I can't bend down that far anymore."

"My hand *and* my knee," I said. I stood up and lifted my leg to the sink, and she pressed her lips against the middle of my knee, exaggerating the sound and time a simple kiss required.

"No more," she said, replacing the lipstick she had lost.

I bent my knee and straightened it several times, making my mother's mouth wink at me. When she was satisfied with her face, she lowered herself onto the side of the tub and sat next to me, pulling my body to hers.

"You're not going to be my baby much longer," she said, resting her chin on my head.

"Because I'm too big, or because you're going to have another baby?"

"Both," she said.

I played with the skin on my knee so that her mouth shriveled up.

"This is what your lips will look like when you're old," I told her.

She pulled my knee to her mouth and planted a ring of red kisses around it.

"I think something special may happen tonight," she whispered.

"The baby?"

"Just something," she said.

"What?"

She smiled. "Let's just see if it happens. I have a feeling."

"What kind of feeling?" I asked, putting my hand on her stomach.

"A special one," she said.

She smoothed my hair with her hand, and I lay with my head on her breast, listening to her heart beat.

"Michael says women sometimes die having babies."

"Michael reads too much," she said, and, after a moment, "What else does Michael say?"

"That babies like openings."

"Hmm. What kind of openings?"

"Boxes. Places to crawl in and out of."

"Michael knows a lot about babies, doesn't he?"

"I guess," I said. "Did I like openings when I was a baby?"

"Not that I noticed."

"Grandma says you liked to read magazines when you were a baby," I said. "She had a basket of magazines by her bed, and you'd crawl over to it and pick out a different magazine every morning and turn the pages without ever tearing any."

"I did?"

"Uh-huh. You liked *Harper's Bazaar* the best."

"When did she tell you that?"

"When I was selling magazine subscriptions for school. She said, 'You remind me of your mother.'"

"I can't believe my mother would remember that about me."

"Why not?"

"I don't know. She never tells me much about myself."

"You were a good baby," I said, patting her arm.

She rubbed my nose. "Give me a kiss goodbye," she said, and I kissed her on the lips, removing the little color that remained there.

"I think Mom's going to have the baby tonight," I told Michael and Mary while we waited for my father.

"She's not due," Mary said.

"She's not nearly due," Michael added. "Besides, Dad wouldn't be taking us to the movies if Mom was thinking of having the baby tonight."

"It's not something you *think* of doing," Mary said.

"Well, whatever," Michael said. "We wouldn't be going to the movies."

"She told me she had a special feeling."

"About what?" Michael asked.

"About tonight."

"It's too early," Mary said.

"She left without any lipstick," I said.

"She did not," Michael said.

"She'd never do that," Mary said.

I offered my leg as evidence.

"So what," Michael said. "She put more on when you weren't looking."

I shook my head. "I saw her leave."

"She'd never do that," Mary said again.

"You probably made those marks," Michael said. "You're always kissing yourself."

"I bet we have a baby," I said.

"I don't bet," Mary said.

"They wouldn't let her have the baby without Dad being there," Michael maintained.

They went out and waited on the front porch while my father changed his clothes. I lay on the living room floor and listened to my father's radio playing down the hall. When it stopped, he appeared above me and said, "Let's go."

My father took us to the drive-in in my mother's car. He didn't like to take his company car to the drive-in, he said, because the gravel that covered the parking lot to keep people from speeding turned his tires white and coated the car with dust. My father always cursed the gravel, no matter whose car we were in, but the sound of it, and the slowness it demanded, made arriving at the drive-in seem ceremonial to me.

We brought our pillows with us, and my father let us take off our shirts and lie on the hood of the car with our backs propped against the windshield. It felt as if we were lying together in bed. My father set the speaker on top of the car, and then he sat down in a lawn chair near the headlights. After three days of rain it was hot and humid, and an hour or so into the first movie my father took off his shirt and hung it over the hood ornament. I was lying between Michael and Mary, and I tapped each of them with my foot.

"So?" they both said at the same time.

"He never does that."

"It's hot," Mary said, but I was sure there was something more to it. It was as if my parents' strange behavior that night — my mother leaving without lipstick, my father removing his shirt — signaled the beginning of something new in our lives.

When the lights came up between movies and music began blaring from the concession stand, my father stood up and stretched. His back was thatched from the webbing of the lawn chair, and the texture made his loose flesh look tighter.

"I'm gonna call and see if your mother got home okay," he said, slipping on his shirt.

"I'll come with you," I said, and I put on my shirt and shoes and rolled off the car over Michael.

My father had me stand in line for soda while he went around the corner to the telephone. There were only a few people ahead of me, and my nervousness was divided between hearing the news of my mother and fearing that my father wouldn't return before I reached the window. I was one person away when he showed up and said, "No answer."

"She's at the hospital," I said. "Call the hospital."

"What?"

"She's having the baby."

"Three small root beers," my father said to the man. And to me, "It'll be at least six weeks until the baby comes."

"Sometimes they come early. She told me she had a special feeling tonight."

"She did? She didn't tell me," he said, paying the man.

"Call the hospital," I said.

"Your mother's never been early. Get some straws," he said.

"No answer," I said as I crawled back on the car.

I took my place in the middle, and my father handed us sodas as though he were serving us breakfast in bed.

"Scoot over a little," I said to Mary.

"I'll be off the car," she said.

I adjusted my pillow so the windshield wiper wasn't digging into my back. "Mom didn't answer," I said again.

"That doesn't mean anything," Mary said.

"This may be the last time we ever come to the drive-in," I said.

"Did Dad say that?" Michael said, sitting up. "We didn't do anything."

"No. I said it. You can't bring a baby to the drive-in."

"Yes you can," Michael said. "Besides, it doesn't have to come everywhere with us."

There weren't many cars at the drive-in that night. On the way to the concession stand, my father told me that was because it was a weeknight. Waiting for the second movie to start, I sighed and said to Michael and Mary, "It sure is a weak night," as if it were my own original assessment.

Michael looked up at the stars.

"You just can't see them through all the lights," he said. "It doesn't mean they're not there."

The lights on the lot flashed, signaling the end of intermission, and Michael nudged me with his soda. "Take off your shirt," he said, "the movie's starting." As I was lifting my shirt over my head, the manager's voice sounded through the speaker. "Emergency phone call," he began, and my father jumped up, tipping over the lawn chair. Mary and Michael slid off the car and we all stared at the speaker. The manager said a name that wasn't ours and we settled back into our places. A few minutes later we heard a car going over the gravel and watched two red taillights exit into the darkness beyond the drive-in. When the movie started, I raised my knee to see what was left of my mother's kisses. Some had retained their shape and some were smeared. Michael put his hand on my leg and gently lowered my knee. "Can't see," he said.

I was thinking about what it would feel like to have a baby crawling over us and between us, the three of us side by side like a new and challenging terrain, when the rain came down without any warning, sending all the people scurrying into their cars. We squeezed into the front seat. My father turned on the windshield wipers, and we bent our heads low and tried to make out the final scenes of the movie, the actors appearing and disappearing before us like a distant mirage. Just as we were getting accustomed to this, the screen went blank. Over the speaker, the manager explained that he could not offer refunds but would be happy to tell us how the movie ended. Immediately, horns began honking, and the manager apologized, saying that the policies, like the circumstances, were sent down from above. A pause followed, filled with static from the speaker, and as my father was about to start the car, the manager cleared his throat and said, "Please. Let me tell you how the picture ends," and he narrated the rest of the film, vividly, to the few cars left on the lot. My father said the manager was better than the movie, and we drove home, all of us still in the front seat, I sitting half on Michael's lap and half on Mary's.

When we turned into our driveway, we could see my mother in the living room window, bending over a playpen that had not been there when we left.

"What the hell?" my father said, barely turning off the car before he got out.

It was after midnight, and we ran across the yard in the rain, entering the house through the front door. My mother looked up, surprised to see us come in that way, and said, "Your feet."

We pulled off our shoes and dropped them on one corner of the carpet. My mother was in the middle of the room, surrounded by baby equipment. Next to the playpen an empty baby swing swung back and forth. My mother laughed and placed a stuffed elephant in the seat. Next to it was something she called a car bed, and then a stroller and a round walker like a circular highchair that sat squat to the ground. She picked up a clear plastic ball from the corner of the playpen and rolled it toward us. When it stopped at his feet, Michael picked it up, palming it like a basketball, and we watched the family of ducks inside fall one way and then the other.

"It's new," Michael said, inspecting it, as if to make certain that it wasn't the same ball he had fished out of the creek the night before.

"Of course it's new," my mother said. "My friends gave me a surprise shower."

"What?" I asked. I was confused by the word and what I thought was some connection to the weather.

My mother explained and reminded me that she had thought something special might happen that night.

"I never heard of a shower," I said.

"I've heard of it," Mary said. "She thought you were going to have the baby tonight," she told my mother.

It was late when we finished looking at the presents. My father had gone on to bed, and we ended the night by telling my mother how he had taken off his shirt at the drive-in.

"He would never do that," she said.

"He did," we insisted, in unison, from where we were lying on the floor. That made me remember my mother's lips, and I was prepared to point to her mouth, to prove to Michael and Mary that what I had told them was true, that she had gone out that night wearing no lipstick. But her lips were perfectly red again. I looked at my leg. The rain had washed off what was left of her there, and I felt an odd sort of loss, as if my mother's willingness to appear in public without lipstick symbolized the special love she felt only for me.

I wanted then to betray her in turn, and I stood up intending to say that I had made up the story I had told her earlier, when we were sitting on the bathtub, talking. I was going to say that as far as I knew, her mother never had a single magazine in her house. But I looked at my mother and she looked at me, and I felt sorry that I had lied to her in the first place, something I was becoming more and more adept at doing. My father said I was a master at telling stories, but I sensed that the stories I made up were becoming dangerous because they filled a need in my mother, and I spun them to fill that need — whatever it was, want of love or release from worry — in increasingly meticulous detail.

Years later, when we were in high school, Mary and I would come home one day to find my mother lying on the floor of her bedroom, slipping in and out of consciousness, coming through only long enough to say, "It's not like my father. Make sure they know it's not like my father." Then she went into convulsions and passed out altogether. She had been depressed, didn't know what was making her so physically ill that day, and was afraid that given her history of depression, people would assume she had attempted suicide. In fact, she had suffered a near-fatal depletion of potassium, and when she regained consciousness, days later, she had lost much of her memory — a loss that would remain, for the most part, permanent. Within a week, she could recall the essential facts of her life: who she was, who we were, where we lived, the major events and circumstances that defined her. But those memories that make intimate our existence in the world, the details of our daily lives, were no longer hers, though some of them would return over time, slowly and sporadically, sparked by a smell or a sound or some other sensation.

It was during this time — in the early days of her memory loss — that my guilt over lying to my mother was at once alleviated and exacerbated. She could no longer remember anything I had told her; we were both free of my fabrications. Yet deep inside, I felt responsible for her illness, for having created a situation — a life of stories far from the truth — that could be resolved only by a sickness so severe that it erased her mem-

ory. Partly, this was exaggerated. I often enlarged my impor-
tance in my mother's life, and, truth be told, most of my stories
were innocent and more than likely meant nothing to her.
But they were a way I could manipulate my mother, telling her
stories that might keep the sadness away if I saw it coming, if I
was as vigilant with her as she was with us, watching our faces
when we slept. This makes it sound as if I were looking out
mostly for my mother. The truth is that though I cared about
saving my mother from her sadness, I cared more about saving
myself from it, from the consequences that came when it covered our house. I was sure this meant that I was flawed somehow, that I was too selfish to love or be loved, and the stories
became like a charm, a means of hiding this fact from my
mother, making her love me more than she might have, more,
certainly, than I deserved. And so, at a young age, I began
composing a life both to console and to fool my mother, to
make her love me despite my faults, and when the shame I felt
in that became too large, and the line between what was true
in my stories and what wasn't became less and less distinct, her
memory of them was suddenly gone, like a life taken merci-
lessly away from us.

That night, though, when I stood up in the living room,
ready to tell my mother I had made up the story about the
magazines, it was spite, not repentance, that drove me to confess. But when I looked at my mother's face, even with the lipstick, I didn't have the heart to hurt her. So instead of confess-
ing, I said goodnight, kissed her, and went to bed. And shortly
after, when Mary joined me, I told her the story about my
mother and the magazines as if telling it to someone else
would make it true, and before she fell asleep Mary said,
"Grandma always tells you more than she tells me." And for
the first time I began to consider the possibility that the stories
other people told were not true, beginning with my father and
ending with the manager of the drive-in, who was in an enviable position, I realized, to make the movie come out any way
he wanted.

As much as I remember the events of that summer, I recall very little about the day my brother was born. It was a Sunday, October 2, the Feast Day of the Guardian Angels, and I remember my father calling in the evening to tell us it was a boy, but nothing else about that day remains with me. The day after his birth, we learned Sean's name, and this too seemed like something that happened far away from us, coming to us as another message over the phone, registering as a strange word we'd never heard.

We sat on the porch saying it to each other. Whose idea was it? we wondered. If it was a boy, my mother had told us his name would be Brian. If it was a girl, Something Mary, since as a pious teenager my mother had made a vow to the Virgin Mary that if God gave her girls, she would name them all in her honor. Sean was never submitted as a possibility. We weren't sure how to spell it, and we were shy about telling anyone the name until the man next door, working on his car and hearing our concern, yelled over, "Sure. Sean. S-E-A-N. Like Sean Connery. You know, James Bond." Only then did we boast about it.

At the end of the week, we went with my father to bring my mother and the baby home, and we were surprised by the large bruise that covered half our brother's forehead. Describing him to us, neither my father nor my mother had mentioned it. Seeing the bruise, I remembered my belly flops and the impact of water on my own body, and I wondered if the water my mother had said surrounded the baby as it grew inside her might have caused too much pressure in one spot. But my mother assured us that the bruise was nothing, that babies sometimes had a difficult time being born, and that within a few days it would disappear. She rubbed her hand over Sean's forehead and kissed the place where he was bruised, and then one by one we kissed our brother, and as we bent down, my mother told him our names, and when he knew who we were, we went home.

· 1982 ·
KANSAS CITY

I spent the summer of 1982 — the summer after Sean died — collecting souvenirs, a small assemblage, really, of giveaways and stolen goods. The first and most functional of these was from a trip we took to Kansas for my cousin's wedding. The room my mother, my father, Kelly, and I shared at the Squire's Inn Motor Motel overlooked a pool and a pancake house, where a round table of our relatives gathered each morning for the breakfast special. *No one could be happier,* it said on the five flavors of syrup stuck like centerpieces on the middle of every table. "I could be happier," my mother said. As if they were exotic elixirs, she tried a different syrup every day. Friday, Saturday, Sunday, Monday. She'd do without the maple.

It was Memorial Day weekend, and the motel pool had just opened for the summer. At night, a reflection of the HAPPY HOUSE PANCAKES sign floated on the blue water, and when Kelly dove in, the red letters rippled but remained legible, like an echo shimmering across the surface. Alone in the pool, Kelly pushed off the side and swam back through the letters, parting the light with her twelve-year-old body. She was the only child among the many aunts, uncles, and cousins who occupied most of the rooms at the Squire's Inn that weekend. I watched her from our second-floor window while my parents got ready to tour Kansas City in my great-uncle's Cadillac.

"I'll leave you the car keys," my father said as he combed his hair in the bathroom mirror.

"We'll probably just get a pizza and watch TV," I said. Directly across from the pool, on the other side of the highway, lines of brightly colored plastic flags fanned out over the parking lot of a Pizza Hut. Strung from every lamppost, they converged at a central point on the Pizza Hut's roof, creating the suggestion of an arrow leading people to the door. They reminded me of the grand opening held at a grocery store near our house when I was little. The signs on the store window promised all kinds of free products, and I read them as I crossed the parking lot with my parents, the dry, sparse sound of the plastic pennants flapping above us. FREE PEPSI WITH PURCHASE. FREE TWO-PAK PAPER TOWELS WITH PURCHASE. "If you have to buy something to get it, then it's not really free, is it?" I said, thinking out loud. I had only recently learned to read, and my parents stopped and beamed at me, but almost immediately I sensed that it wasn't just my reading that made them proud. I had figured out something about the value of a dollar, and from their response, I understood in an instant what was important to them.

Now my father joined me at the motel window. "Pizza Hut? Think that'll fly with your sister?"

"Hard to say," I said.

"Well, drive over there if you go," he told me. "Or have them deliver."

Kelly was standing at the side of the pool. With her hair wet and the night all around her, she looked smaller and more alone.

"Is there a lifeguard down there?" my mother asked, coming in from her brother's room with a bucket of ice.

Back in the pool, Kelly was floating behind a beach ball so big it blocked her head from view.

"She's fine," I said. The flowers of her bathing suit, a patch of pink and green, glided along like an oasis under water, her leg breaking the surface occasionally with a listless kick.

My mother looked out and saw there was no lifeguard.

"Isn't that against the law?" she asked.

I pointed to a sign on the fence. In letters large enough for us to read from our window, it said that children were not allowed in the pool without a parent.

"Go down there then," she said.

At the pool, I rolled my jeans to my knees and dangled my feet in the shallow end.

"Come in," Kelly said. The cars on the highway created a steady stream of sound behind her, darting and buzzing in both directions. "Come in," she said again, swimming away from me.

"I didn't bring a suit."

"Didn't you ever skinny-dip?" she asked in a voice both challenging and curious.

I watched her float past me. Her legs, lanky and girlish in dry daylight, looked less innocent under water at night. It was as if the refraction from the water and the pool's lights transformed her coltlike limbs into legs that were shapelier, more womanly. Watching her, I realized I remembered nothing about the development of my own body — the beginning of breasts, the appearance of hair under my arms and elsewhere. I did remember shaving my legs under Mary's direction, locking the bathroom door against intruders, against the danger — and this on a night when no one else was home — of getting caught in the act. I hadn't asked my mother for permission, and though I knew she wouldn't disapprove — I was old enough, the same age as Mary when she started, the same as Kelly, who would shave her legs for the first time the next morning, refusing to attend my cousin's wedding otherwise — I knew also that it was the kind of occasion in our lives that my mother always wanted to witness, and that by doing it alone, with Mary, I was taking something away from her, something that made us mother and daughter, one of the milestones she might have had in mind when she prayed to God to give her girls. I didn't deny my mother this maliciously. I had simply reached that urgent moment — I *had* to shave my legs — on a night she happened not to be home. It was summer, of course. We locked the door and whispered.

The next night, I went with my mother to buy dinner at a drive-thru, and as we waited in the line of cars wending their way to the pick-up window, I said, "I shaved my legs."

"When?"

"Last night."

My mother reached over and ran her hand up my leg.

"I wanted you to wait a little longer," she said. "I liked your legs."

"But I wanted to do it. Look, no nicks," I said, placing my leg in her lap.

"Did you use a razor?"

"Yeah," I said. "I was really good at it. I didn't cut myself anywhere." I held up my other leg to show her. "I did it dry, too," I added, "without any soap and water." I believed this last fact established my skill at shaving, and I had been waiting all day to say so. I wanted to ask my mother if she bled at all her first time, but in the few seconds it took me to form the question, she looked right at me and said, "You used a *real* razor?"

"Yeah. Mary showed me."

"Where'd Mary get a razor?"

"You gave it to her. Last year."

"I never gave Mary a razor," she said, examining my legs more closely.

"Yes, you did. For Christmas. The Lady Electric."

She laughed and moved the car up in line.

"What's funny?" I said.

"You *can't* cut yourself with an electric shaver."

"You can't?"

"No. Razors are what cut you. Like the kind your father uses."

"Are you sure?"

"Yes," she said.

"I thought I was doing the real thing. The dangerous thing," I said, feeling deflated.

When we reached the window, my mother handed me a bag filled with four chocolate shakes, and I held it between my knees. The cold was already beading up on the cups, and the bag was getting wet. Soon a trickle of water ran down my thigh. I put the bag on the floor, steadying it with my feet. The concentration it required made my legs rigid. Smooth now, they looked younger and older at the same time. I was suddenly sorry I had shaved them.

The bags of hamburgers and fries felt warm against my skin. On the way home, my mother took my hand in hers and kissed it. "Why do you always want to do what's dangerous?" she asked.

"I don't know," I said.

"Your legs look nice," she said, sensing my disappointment, the false thrill I'd had in not hurting myself. "But now you'll have to keep it up for the rest of your life," she said, and she sighed, as if she were sorry the rest of my life had to start so soon.

"Do you want your own Lady Electric?" she asked before we turned down our block.

"Not if it's going to count as a present," I answered. "Anyway, wouldn't a real razor be cheaper?"

"No razors," she said as we pulled in the driveway. She stopped the car and turned my face toward hers. "I'm not ready for you girls to be bleeding like that," she said.

Michael had come out, ready for the food, but he stopped short when he saw her holding my chin. Still smooth-skinned at sixteen, he hadn't yet started to shave.

"What about Michael?" I asked. I was prepared to argue the point if my mother planned to buy my brother a real razor when he was ready.

"Lady Electric," she said, and we laughed, and Michael grinned at us from the lawn, wondering whether it was okay to come closer.

Floating in front of me in the motel pool, Kelly extended her arms above her head and spread them in semicircles down to her sides, sending a wave of water toward me. Before I could pull my feet out and stand up, my jeans were soaked.

"Watch it," I said.

"You didn't answer me," she said, swimming away again.

"About what?"

"About skinny-dipping."

"Oh. No, I never have."

"Why not?"

"I don't know," I said. I was twenty-four and hadn't gone swimming since my sophomore year in high school, when a three-week swimming session in gym class was mandatory. The night before it started, I jumped off the roof of our garage a few times, hoping to break an arm or a leg so that I'd be excused. Nothing in my body would break, and though it occurred to me, I was too scared to jump off the roof of our house, two stories tall. So I endured the three weeks of swimming, holding my breath against the embarrassment of my body, so much bigger than the bodies of my classmates. I wanted to die every day.

"I do," Kelly said, not prepared to give up the topic. I thought at first she was trying to tell me that she knew why I didn't skinny-dip, but she meant that she herself had skinny-dipped, was a regular skinny-dipper, believed skinny-dipping to be a secret thing that made her daring.

"I do it all the time," she said.

"Where?"

"Wherever."

A car honked in the parking lot, and six adults waved from my great-uncle's Cadillac. I noticed that my father was at the wheel, my uncle, near eighty, in the seat beside him.

"What are we going to do tonight?" Kelly asked.

"What do you want to do?"

"What time does the pool close?"

"I think it closes whenever you quit swimming," I said.

I thought that if Sean were here, swimming with her, they would probably play a game of Marco Polo, and I imagined the sound of their voices, Sean's newly deep and teasing, Kelly's more direct. "Marco," Kelly would call out. "Polo," he would answer, already going under, swimming silently to the other side of the pool to lead her astray. MAR-coh. PO-loh. The o's sounding longer and slower, as if they were written in water. And then she would find him, catch his foot or his shoulder, ending their verbal blind man's bluff, and they would decide whether they wanted to do it again, the two of them alone with the words and the water and nothing else but the night.

"We could get a pizza," I said.

"What kind?"

"Pizza Hut."

"But what *kind*?"

"Whatever kind you want."

"I only like hamburger," she said.

"I think that's well known," I said. As a family, we had long ago conformed to Kelly's tastes. She had a short list of what she would eat, a long list of what she wouldn't. She was the most stubborn and headstrong among us, the first to demand things from my mother — her own food, for instance, if dinner that night was something she didn't like. And my mother responded as if she had been waiting all her life for one of us to want her in that way. They fought and screamed at each other, rarely kissed or caressed. It was a relentless relationship, a war they were both winning, and to witness it was to realize that Kelly was alive in a way the rest of us weren't.

"Then what?"

"Then what what?"

"Then what'll we do?"

"Watch TV," I said.

"Wow," she said.

I fingered my father's car keys in my pocket.

"Here," I said, tossing them into the pool.

"What's that?"

"Sea Hunt," I said.

"What?"

"Didn't you ever play Sea Hunt?"

"Yeah. So?"

"So find the car keys at the bottom of the pool."

"I'm not finding them."

"Kelly."

"No. You threw them in."

"Well, we can't go get pizza then."

"So what. They'll deliver."

"Get the keys," I said.

"You get them," she said. "I hate Sea Hunt."

She got out of the pool and started drying herself.

"I have my own room key," she said, retrieving it from where she had rolled it up in her towel. "Mom gave me hers."

I watched her walk across the parking lot and go in the side entrance of the motel. A few seconds later, the light in our room went on and she walked past the window, toweling her hair. A few minutes more, and the blue light of the TV began blinking.

I lay down beside the pool. The close smell of chlorine and the hard certainty of the cement felt satisfying and familiar. As people left the pancake house, their voices, their footsteps, the sound of their cars, their laughter sometimes, came to me in a sequence that was orderly and right. In the scene around me, I was the only interruption.

The car keys lay somewhere in the deep end. Above me, the stars were low in the sky. I closed my eyes and listened to the voices and the cars and the water, like a gentle sleeping body beside me, massive and undisturbed.

"I'm sleeping out," he would say.

"I'm sleeping out too," Kelly would say, following him to the tent that Michael had pitched for them in the back yard. My mother would leave the kitchen door unlocked, telling them to come in if they got cold. Kelly in her baby-doll pajamas, Sean fully dressed in his shirt, shorts, shoes, and socks. They took blankets and flashlights and books. Sean set an alarm clock in the corner, and they discussed what they would eat for breakfast. It had to be something special. It couldn't be cereal after a night spent sleeping out. They would ride their bikes to the doughnut store at daybreak. Or sooner, Sean said, resetting the clock. Some nights, I would sneak out and scare them, or Michael and I, creating a sharp and sudden clamor. Other nights, I'd lie near the tent and listen. "What's that?" Kelly would say, and when they were sure it was nothing, they'd resume their inquiry into each other's lives. They were six and nine now, or eight and eleven. The questions were easily asked and easily answered. Are you going to smoke when you grow up? Would you rather listen to records or the radio? Which did you like better, first grade or second? Are you scared? Are you cold? Are you going to go in? If I go, will you go? It wasn't the answers or the asking. It wasn't the need to learn something new. It was the night watch, the act of being alive, of being innocent and equal, as Michael and Mary and I once were, open and uninhibited, unaware that anything would ever come between us, that our bodies would develop, leading us away from our ignorance, our innocence, usurping our souls, separating us from one another, leaving us too self-conscious to reveal what we would once have revealed, just days ago, months ago, just hours earlier.

And so I would lie quietly and listen, scooting as close to the canvas wall as I could, not wanting to scare or startle them, not wanting to impose on their privacy, but yes, imposing on it; not wanting to discover their secrets, but yes, discovering them; not wanting to love them more or love them less, but yes, loving them, listening to them, liking them, wanting to be them, beside them, bodiless, wishing that love, all love, could remain whole and unexpressed.

Kelly would be back in her bed by morning, and Sean, ever determined, would wipe the dew off his bike and go for doughnuts. It was the way they were. It was the way we all were, defined as much by ourselves as by each other. I being unlike Mary. Michael being more or less like me. Sean being more diligent day by day than Kelly. Kelly outliving him in the end.

"If I go, will you go?" she would ask him. Afraid of the dark, she didn't want to enter the house by herself, climb the stairs to the second story, and listen to her fears — noises, shadows, and other uncertainties — as she waited to fall asleep. "If I go, will you go?" she would ask, and always he would accompany her in, going back out later alone.

So I saw, as I lay under the stars in Kansas, that it was Kelly, not Michael or Mary or I, who was mourning him most. It was Kelly, left to live out her childhood without him, he who was her childhood, as Michael and Mary were mine.

"They don't deliver," she said as she dove in. I had fallen asleep, and the sound of the water, more than her words, woke me.

"Did you see where they went in?" she asked, surfacing.

"Under the first *p* in Happy House," I said, but as I stood up, I saw that the restaurant's sign was no longer lit, the logo no longer rippling across the water. "What time is it?"

"Ten after ten," she said. "C'mon. They close in twenty minutes, and they don't deliver."

"What do you want me to do?"

"Help me find the keys!"

I stood at the point where I was standing when I threw them into the pool.

"Can you see this?" I said, holding up a quarter.

"Kind of."

"Well, watch where it goes in," I said, and I threw it, aiming for the place where the *p* had been.

Three times, Kelly went under and surfaced without success, retrieving neither the coin nor the car keys.

"Forget it," I said. "We'll walk over."

We stood at the side of the highway, waiting for a pause between cars. "I hope they don't close," she said. She looked at the waterproof watch she wore. It was Sean's, and because she was slender, it looked like a large weight on her wrist.

"Ten-twenty," she said. Her towel was draped over her shoulders, and her hair was dripping.

"You should stay here," I said, looking down at her rubber thongs.

"I can run in these," she said.

"No, wait here," I told her. "Wait in the chairs by the pool."

"No. I can run in these," and she began running in place,

not to prove that the thongs were safe but because she was worried we wouldn't get across in time. "We just have to make it to the median and then over again."

"All right." I saw a long gap in the stream of cars. "Ready?" I asked.

She nodded, and as the last car zoomed by, she took my hand, letting go when we reached the median, taking it again when we ran to the other side.

Even though there were still people in the Pizza Hut, the man seemed annoyed when we ordered a whole hamburger pizza. "That'll have to be to go," he said. "Anything else?"

"Let's get one of those," Kelly said, pointing to a pyramid of red-and-white picnic jugs at the end of the counter. She picked up one by the handle and swung it back and forth. "Free refills all summer," she said, reading the sign.

"Do you have to get these refilled here?" she asked the man as he filled the jug with Coke. "Or can you take them to any Pizza Hut?"

"Any Pizza Hut in the forty-eight contiguous states," the man said, sighing behind the register and shifting his feet in a way that made plain his impatience.

The lights in the parking lot were already turned off when he unlocked the door to let us out. He held it open with one arm, and we walked past him as if we were being released.

"Man," Kelly said when he locked the door behind us.

As we stood on the side of the highway, she held the picnic jug between us, and I felt the cold of it against my leg.

"Now," I said, and we ran to the median.

The traffic was heavier than when we'd crossed before, and we stood on the median for a long time without talking. Though headlights were everywhere around us, it was as if we were standing in total darkness.

"What's a contiguous state?" Kelly asked after some time had passed.

"A state that shares a boundary with another state," I said. "He meant you can get it refilled in any state except Alaska and Hawaii."

"I didn't think they had Pizza Huts in Hawaii."

"Did you think they had them in Alaska?"

"Well, they don't have anything in Alaska," she said, "but it doesn't seem like there should be Pizza Huts in Hawaii."

I knew what she meant. There was a restrictive geography to fast food. Its presence in some places, or in our idea of those places, seemed odd and unsettling.

"Maybe they don't have them there," I said.

"Maybe not." She set the Coke down and shook her arm.

"What are you going to wear to the wedding?" she asked.

"A dress," I said. I looked at her, the thinness of her, her dark hair almost dry now, her toenails painted pink. "That green-and-white dress," I added. I thought of the difficulty I'd had in finding a dress, how I almost hadn't come to Kansas, until my mother happened on a halfway stylish dress at a store that carried large sizes and convinced me that it looked fine. I wasn't eager to wear it or to go to the wedding.

"Are you going to dance at the wedding?" Kelly asked.

"I doubt it."

"Not even with Dad?"

"Maybe with Dad. If he dances."

"Why wouldn't he dance?" she asked. "He always dances."

"I guess he will then," I said. I thought of the way he stood stiffly in the background now at most family functions, observing events from what seemed to be a great distance. I thought of the way he smiled now, the painful smile he had assumed since Sean died. It was like watching a shy child smiling on command before a camera. I couldn't look at him anymore when he smiled. I couldn't bear the fragility of his face. There had been no occasion since Sean died that called for dancing, and I couldn't help but wonder if my father's dancing style — he had been smooth and unselfconscious — would be affected by grief in the same manner, turning what had been effortless into something strenuous and difficult to sustain. I didn't say any of this to Kelly, just as I hadn't explained any of the details surrounding my dress.

"Do you think anyone will ask me to dance?" she said.

The question stung me. It echoed my own anxiety and exposed a vulnerability that we seldom saw in Kelly. I imagined
us sitting together at the table while everyone else was on the
dance floor. I was the only one of my cousins unmarried; there
would be no one at the wedding Kelly's age. Uncles would ask
us to dance out of obligation. Cousins would goad their husbands into taking us out on the floor. Kelly would be tickled by
this. I would find it awkward. She would revel in the line and
circle dances, delighted that she could join in on her own, and
I would wait for her, wearing my father's smile.

"Of course someone will ask you to dance," I said. "Dad
will. Michael will."

"I mean besides them."

"Uncle Harry and Uncle Walter will," I said, and as I peered
down the highway for a hole in the traffic, I listed all her potential partners.

"Is it really true," she asked, "that you and Michael won a
Twist contest when you were kids?"

I smiled. "Yes," I said. We were at the Excelsior Springs Hotel, in Excelsior Springs, Missouri, on our first family vacation.
I was five or six, Michael nine or ten. Mary wouldn't join us
on the dance floor when, after dinner, the conductor of the
band announced the contest. It was only for kids. "C'mon,"
Michael said, grabbing my hand. We both knew the Twist.
The band played a Chubby Checker song, and Michael danced
as if he were truly transported by the music and the Twist. He
was tall and skinny, and the movements looked as though they
had been made for his body. And I did him no harm. Though I
didn't cut the same sharp angles, there was an odd symmetry
between us, and a noticeable happiness. It was one of the purest moments in my life, maybe in his too. We were good, and
we knew we were winning.

A few years after my cousin's wedding in Kansas, the same
configuration of family would reconvene at the Excelsior
Springs Hotel, site of our first family vacation, to witness Michael's marriage. It wasn't nostalgia that caused him to hold

his wedding at the hotel; Excelsior Springs was his wife's home town. But the coincidence delighted him, and for weeks before the date he threatened me with the Twist.

I looked down the highway. The traffic showed no sign of letting up.

"What'd you win?" Kelly asked after I told her, in great detail, about dancing the Twist that day.

"I don't remember," I said, but as soon as I said it, the gold beach balls came back to me. They were huge plastic balls with sparkling glitter inside. Michael gave his to Mary; it matched her sunglasses. But the real prize was walking back to the table — the crowd clapping, our parents beaming with pride.

"Have you ever won anything else?" she asked. And I told her about every prize I had laid claim to. We were in a safe place, the past, and it was easy to tell her about things that had happened before she was born. From the beginning, she had been eager to hear every story, as if, born last in our family, she had arrived, an immigrant, in a country that had already been settled. Regarding our past as a territory she was destined to discover, she hoarded every detail. And if we were not forthcoming, if something came as a surprise to her, she would put her hands firmly on her hips, a stern and serious six- or seven- or eight-year-old, and say, "Why don't I know about this?" She was vigilant and unwavering. "Why don't I know about this?" she wondered when she learned that my father used to go on weekend religious retreats. After keeping a vow of silence for three days, he returned home looking spiritual but shorter, I thought. "Why don't I know about this?" she wondered when it came out that my mother had gone to college. Why didn't she know that Mary broke her foot the summer after fifth grade, that Michael was once beaten up by a boy down the street for wearing a yellow shirt with big blue polka dots? The boy said it was sissy. It was the first shirt Michael had ever picked out for himself, and he defended it with his life, destroying it in the battle, crying over it when he came home. Why didn't she know that my father lost his job on the day she was born, that my mother once dyed her hair navy blue, that Sean was born bruised? Why hadn't we told her these things? What were we waiting for?

We'd been on the median long enough for the pizza box to turn from hot to warm against my hands. Kelly had picked up the picnic jug and put it down again, her arm tired from the weight of it. The flow of cars was constant, but we were no longer anxious about getting across. Kelly asked one question after another, questions at once significant and insignificant, each aimed at illuminating some aspect of the past, leading to some anecdote she had not yet uncovered. And sometimes the answers fell forward uncontainably into the present, revealing a feeling instead of a fact, causing us to recognize ourselves in each other, to say something that had no boundary, something that was not part of the past but part of the country ahead of us, the one we could arrive at together, immigrants to a place that still had possibilities.

Cars passed in waves all around us. The sound of their radios spilled out like molecules of music, elongated songs streaking unrecognizably by. Drivers honked. Boys shouted. Heads leaned out at us. Yet we were alone on our island, the Pizza Hut of Hawaii, alone in our own acceptance and understanding, alone in a lapse that wouldn't last.

And then the line of cars began to thin. I touched Kelly's arm to alert her. She was asking me something about seventh grade.

"Ready?" I said. "Now."

We stepped off at the same time, but when I reached the other side, Kelly wasn't with me. Just as I looked back, a car ran over her thong, sending the thin flap of blue rubber tumbling down the highway. I instinctively ran toward it, but I couldn't reach it. More cars were coming. I scanned both sides of the highway, but other than the thong, there was no sign of her. Where was she? Where the hell was she?

I was standing in front of a row of hedges between the motel pool and the highway. Even over the hamburger pizza, I could smell the chlorine behind me, and the combination, along with the speed of the cars, made me sick to my stomach. My sister was nowhere in sight.

Wouldn't I have heard her scream? Wouldn't I have heard cars skid and honk? Her thong became airborne again and again, as one car after another lifted it farther down the highway. Where was she? How could she suddenly not be here? How could she disappear?

I was holding the pizza box as if it were a precious burden, my arms bent before me. I hadn't given her a hand to hold on to. "Kelly!" I shouted, but the cars absorbed my scream. The sound of my voice, the urgency of it, would never reach the other side. If she was on the other side. "Kelly!" I yelled again. I thought of the car keys at the bottom of the pool. "Drive over there," my father had said, "or have them deliver." "I can run in these," she'd said. "Kelly!" I was crying now. "Come here!" I pleaded, as if the force of that command, issued into the air, would bring her back.

"Come home!" I cried when Sean died. "Please come home!" I would be driving my car, and the terror of his absence would come over me. "Come home!" I would cry. I drove and drove after he died. I couldn't stop. Off work at midnight, I drove until daybreak, highway to highway around St. Louis, going nowhere, crying that command. "Come home!" And I half believed he could. If he could disappear so quickly, I was certain, somewhere in my mind, in my heart, that he could just as quickly reappear. I'd been brought up to believe in resurrection, so why not believe it? Why not push for it? Why wait for the end of the world? My mother believed in miracles. This was the only one I wanted. But I dared not tell her. As she prayed for Sean to find peace in heaven, I prayed for him to come home. I dared not tell her. I dared not tell anyone. I drove from midnight to dawn every day. The world was white with winter. It was ready for warmth. It was ready to receive him. He could slip back in. I would make the world — our family, his friends — forget that he'd been gone. I would smooth it over for him, fill him in on time past, help him to live again. I would welcome him home.

Years have passed since I drove from midnight to dawn every night, and still I would do this for him. Sometimes, more seldom now, more calmly, no longer raging into the darkest night — maybe I'm washing the dishes after dinner, maybe I'm walking down the street — I say, "Come home, Sean." I am still waiting, and I would not be surprised if he suddenly showed up. It would be something I could easily accept. Because of him, I believe in miracles. Maybe it is madness. Maybe, beyond the laws of nature, it is love.

"Kelly!" I yelled. But there was only the incessant stream of cars. I was crying, and my tears added to the blur everything was becoming. "Kelly!" Could she have been snatched? Could a car have slowed down enough for someone to pull her in? No. It wasn't possible. I would have heard it. Wouldn't I have heard it? Wouldn't she have screamed? She wouldn't have gone quietly, would she? Maybe someone had been on the median with us, farther down in the dark. Someone we hadn't noticed as we talked the time away. Still, she would have called out to me. Wouldn't I have heard her? If she was in danger, wouldn't she have called my name? "Kelly!"

I didn't know what to do, how to find her. "Kelly!" I screamed, and I knelt on the side of the highway and cried. Cars hit me with their headlights. I curled over and shielded my eyes from the glare. Did anyone notice me there? Did they care? Had any of them seen her? Should I start flagging them down? I straightened up, screaming, and then I saw her, like an apparition before me. She was hopping on one foot, stopping, hopping again, inching her way toward me. I ran to her.

"I lost my thong," she said, crying.

I touched her cheek. "Where have you been? Haven't you heard me calling you?"

"I forgot the jug," she said, lifting it a little. "I forgot to pick it up when we stepped off, and when I turned around to get it, I tripped on my thong."

She had fallen, face forward, onto the median. When she stood up, she was confused. She crossed back the wrong way, and when she found herself in front of the Pizza Hut, one foot bare and bleeding, she sat down on the curb and cried.

"I called you!" she screamed.

"I couldn't see you," I said.

"I called you!" she screamed again. "I could've been killed!"

I ran my hand over her cheek and spread my fingers through her bangs, pulling them back so the full oval softness of her face was showing. She was mad at me. She had been mad at me for months. Ever since Sean's death, I had removed myself from her. I couldn't bear to be with her. I knew what it was now. I couldn't bear what was missing between us — *he* was missing between us — and I couldn't be with her without being acutely aware that he wasn't there. It had been like that. It had always been the two of them since the day she was born. There were few photographs of one without the other. They were like their own family. Sean and Kelly. Their names were inseparable. Call Sean and Kelly, my mother would say. Where are Sean and Kelly? Take Sean and Kelly with you. This is Sean and Kelly, I'd say to my friends, as if they were one person.

I wanted to kiss her, but I couldn't. I wanted to pull her close to me and tell her everything I was feeling, but in the months since Sean's death, I had already lost the ability to do that. I wanted to tell her that she didn't need to die, that she didn't even need to come close to dying, for us to recognize that she was still alive. And since I didn't say these things, she had no way of knowing what her life meant to me. She had no idea how scared I was at the thought of losing her, how agonizing those moments had been. And why couldn't I tell her the one thing that she needed to know? I would try again and again that summer but would never succeed, and by the time she reached her thirteenth birthday, six months later, she had lost the energy to draw it out of me, to draw it out of any of us. Instead, I would watch the trauma of her teenage years as she tried over and over, in one furious way or another, to remind us that she was alive. Maybe that night, safe now on the side of the highway, I could have changed who we were both becoming. But I didn't. I couldn't. We had already begun to walk in opposite directions into the same darkness, Kelly taking a loud and angry path, I a silent, inscrutable one. It was as if we had entered a new country, one with no past whatsoever, and in this new country, we would never say his name.

"It's getting cold," I said.

"Is that all you care about? The pizza?"

"No, I mean it's getting cold out here." Still in her swimsuit, she was standing on one foot, shivering, and I saw that her arms were covered with goose bumps. I started to walk toward the motel.

"Help me!" she said.

I took the picnic jug and stooped down in front of her without saying anything. Without saying anything, she put her legs around my waist, her arms around my neck, and I raised myself slowly. She was light, and even at twelve, almost a teenager, she was no burden to carry piggyback, no burden except that her long legs were hard to find a place for. I raised the pizza box so she could cross her legs beneath it. When she was steady, she held my shoulder with one arm and took hold of the jug again, letting it swing beside us. Her breath was warm against my ear, and I could feel the soft beginnings of her breasts pressing my back.

"The pizza *is* cold," I said as we crossed the parking lot, but I knew she preferred her pizza cold, liked it even better left over, eaten right out of the refrigerator in the morning. We had been gone less than an hour. Most of it had been time spent waiting — for the pizza, the traffic, or each other. But I felt completely tired, as if I had been carrying her this way all day.

When we reached the motel entrance, I stopped, and she lowered her feet to the ground. "I like it that way," she said, running ahead of me. She was delighted by the temperature of the pizza, the prospect that it would grow colder. It was one of the things I admired most about her: a demanding child, at times she was easily pleased.

The next morning Kelly woke early, went swimming, and came back with the car keys. I hadn't gotten up yet, and she sat down next to me on the bed and slammed the keys on the nightstand. A spray of water landed on my forehead.

"Can someone drive me to the store?" she asked after a few minutes. I moved away from her; her suit was making a wet spot on the sheets.

"What store?" my mother said from the other bed.

"Any store. I need a few things."

"What things?" my mother said. My father got up and went to the bathroom.

"I'm shaving my legs today," Kelly said. "I need whatever you need to do that. A razor, I guess. And maybe some special lotion. What else will I need?"

"You won't need anything. You're not shaving your legs."

"Yes, I am," Kelly said.

"Do you have to do this today?" my mother said. She was referring not to the shaving but to the scene that was about to start.

"What's wrong with doing it today? The wedding's not until one."

My father came out of the bathroom, dressed. "Let your mother sleep," he said.

"She's slept enough," Kelly said. "Who's going to take me to the store?"

"Kelly!" my mother said. She hadn't moved since Kelly started talking, but now she sat up. "You're not shaving your legs!"

"Well, I'm not going to the wedding this way," Kelly said. "If I can't shave my legs, I'm staying right here, and when everyone asks where I am, you'll just have to tell them how stubborn you are."

"How stubborn *I* am? Tom, will you talk to her?"

"Listen to your mother," my father said. "We're here to have a peaceful weekend."

"I'm not doing anything to disturb that," Kelly insisted.

"You're too young to shave your legs," my mother said. Her

voice was flat and unconvincing. As soon as I heard it, I knew I would be driving Kelly to the store.

"I was almost killed last night on the highway!" Kelly said. "Shaving my legs is nothing compared to what I survived." She had already told my parents the story, and she was ready to capitalize on it now.

"Oh, Kelly," I said. I propped myself up on my elbow and looked at her for the first time that morning. Her hair was uncombed, and a towel was wrapped around her waist.

"I don't think this is your conversation, thank you," she said.

I started laughing. At some point in all her confrontations, she would speak with more formality and forcefulness than her small body could bear. As a three-year-old, she would stop us at every turn if we were hurrying her too much or acting the least bit angry with her. "Be patient, please," she would spout. It was her first phrase of nonnegotiable politeness, terse and to the point, like a language none of us knew.

"Be patient, please," she would say in a quick, no-nonsense tone that reminded me of the pharmacist's mynah bird at the drugstore down the street. "Awkk! Be patient, please." It became the phrase our family adopted to put an end to any argument or heated exchange, a phrase we couldn't hear or say without smiling, conjuring up the memory of Kelly reigning over us from her highchair. "Be patient, please," she would say from whatever perch she was occupying, and we would all take heart, put ourselves in her position, yield, congenially, to her command.

"Be patient, please." I laughed and lay back down. I could see my mother's body shaking underneath her sheet before her face popped up, red with laughter. My father tried hard to hold back a grin as he slipped on his shoes.

"I don't think this is very funny," Kelly said, and we laughed all the harder, until she could no longer take herself seriously and laughed too.

"I *am* shaving my legs," she said when the laughter subsided, but by that time we all knew she would be shaving her legs, because the minute my mother laughed over anything, it was an indication that whoever was arguing with her had already won.

"Use your father's razor then," my mother said. She stood up, slipped a cotton housecoat over her nightgown, and — though we were in a motel with maid service — immediately began to make the bed. It had been a long time since she'd left her bed unmade in the morning, a long time since she had laid back down in the middle of the day. My mother was no longer plagued by depression. For nearly a decade she had been free of it, and even Sean's death did not carry her back to that dark, unendurable place. She was at peace. Though mournful and full of despair, she would somehow remain at peace, spared the depths of her previous depression. Years later she would tell me that she thought the years following her father's death were God's way of preparing her for the loss of Sean. She had suffered ahead of time, so that when Sean was taken from her, she would be better able to bear it.

"Get up." She nudged me as I pulled the covers over my head. I had planned to sleep until it was time to get dressed for the wedding, but she started to make up the bed around me, tickling me as she tucked in the covers.

Just when we thought everything was settled, Kelly said, "I'm not using a *used* razor. I need a new one for my first time."

"Your father's is new enough," my mother said.

"Right," Kelly said. "It's only about forty years old. Anyway, I need something a little more feminine."

It had been a long time since the Lady Electric. I drove Kelly to the store to buy a razor. She chose a bag of ten disposables. They were pink. "These'll be good. You can just throw them away after," she said. Against our advice, she used a new one for each leg, and as I watched her run the razor in short, timid strokes across her skin, I thought about the first time Sean shaved. He was the first of us to start out doing it the real way, the dangerous way. The day before, we had all been together for his fifteenth birthday. My mother made lasagna, and a few of Sean's friends came over. As we walked into the dining room, I looked at his face and saw the fuzz on his chin. Something about the light in the room made it more noticeable. It was soft and blond, but substantial. "Wow, Sean," I said, "you need to shave." I said it abruptly, forgetting that his friends were there. It was a time when everything about him was changing — his voice, his body. Even the features of his face were starting to sharpen into a look that was more handsome than cute. "Let me see," Mary said. "I want to see," Kelly said. Michael took Sean's chin and pointed it toward the light. "Yep. You better get started on that," he said. Sean grinned shyly, pleased and embarrassed.

The next day, without conferring with anyone, he went to the store and bought a solid, silver-colored razor with a replaceable blade. "It's nicely weighted," he said, revealing it to me several days after he'd begun to use it. Like my father, he chose things that would last a lifetime. Along with the razor, he bought a few boxes of blades. He was set, he said. I touched his chin. It felt smooth and masculine. As with most things in his life, he had performed this act privately. He didn't ask for instructions or assistance. If he had asked my mother, would she have objected to his using a razor? No. It was Sean. He was competent and careful. She could always trust him to do the right thing. With Sean, there was never any danger, never any need to worry. With Sean, everything was safe.

As Kelly grew more confident, using longer, smoother strokes on her second leg, I became frightened that she'd hurt herself. The more adept she became at shaving, the more I held my breath against the inevitable nick, the free flow of blood from her body. Watching her, I thought about Sean's wrists, how he had tried to slit them, how he had shown the scratches to my mother, offering them up as evidence of what he had done, as if she would not otherwise believe that he had swallowed handfuls of my father's heart medicine. And he was right. She could not believe it. It was unbelievable. She made him show her the bottle, near empty now. Was it out of consideration that he had left a few pills for my father? Probably. Probably he thought about the time it would take to get a prescription filled. It was the weekend. He would not want to endanger my father's life. How many did you take? my mother asked him. This many, he said. No, it was not just that she couldn't believe it. It was that she wanted to be able to tell the paramedics everything they might need to know to save him. She was smart that way. She was clear-headed, asking Sean everything she thought essential to his salvation.

I never saw the marks, but from my mother's description, they were probably too slight to have turned into scars. Still, his wrists represented a part of his death I couldn't bring myself to think about. Every time the thought came to me, I crossed my arms around my body, as if I were cold, tucking in my wrists, protecting the blue line of life within them, so visible, so vulnerable. I couldn't imagine him hurting himself. It was the part of his death I tried to forget, and now, seeing Kelly shave her legs, guarding myself against the possibility of the blade entering her flesh, of having to help her slow the blood and stop it, I realized that if he could have cut himself, he might have lived. If he had slit his wrists, if he had opened a vein, he might still be here. He might have tried, instinctively, to stop the bleeding, to have shown my mother sooner. And for the first time, as I watched Kelly shaving, I wished he had cut himself. I wished he had sliced right through himself and recognized the wound for what it was.

"Be careful," I said. She was nearly finished, and deciding that she didn't need me any longer, I went into the bedroom and poured myself a Coke. My mother was making up her face in the mirror. "How'd she do?" she asked as she filled in her eyebrows with a brown pencil. While Kelly was in one room removing hair from her body, my mother was in the other adding the illusion of hair to her own. It was the perfect representation of their relationship: they always took opposite approaches to achieve the same end, in this instance an appearance they found presentable and pleasing. Before I could answer, Kelly came out and offered her legs for inspection. My mother ran her hand over one leg and then the other. "Nice," she said.

"Did I miss any?" Kelly asked, making my mother look at her legs from every angle.

I could see the smoothness of her skin from where I was sitting. Her legs looked longer now and more defined, as if that fine layer of softness, something almost unnoticeable, had been keeping them girlish, concealing their true shape. She had not cut herself, not one nick. She had done the dangerous thing. She would do it again and again, surviving the occasional cut. My mother went back to her eyebrows.

"I'm thirsty," Kelly said.

I handed her the picnic jug. It was our first souvenir of the summer.

LOUISVILLE

We had not been back from Kansas for long before Kelly and I took another trip, this time to Kentucky to visit my aunt at the motherhouse of the Loretto nuns. It would be the first time Kelly slept in a convent. I hadn't slept in one since I was a child, when we visited my aunt regularly, traveling to whatever city she happened to be stationed in. In those days, when we drove long distances to see her — it was before Sean and Kelly were born — I felt our excursions fulfilled a religious obligation along with a familial one. It was as if we were on a crusade from convent to convent, and my aunt, rather than being merely my father's younger sister, represented a crucial connection to our Catholic upbringing, to the very core of our spiritual well-being, as if our entire family received God's favor from her vocation. Michael said we were granted an indulgence whenever we visited her. Every time we stepped into a convent, he said, a day was subtracted from the many years our souls were slated to burn in Purgatory. That's why whenever we visited my aunt, Michael would ask to go outside. It was his intention to accrue a multitude of indulgences each day by going in and out of the convent as many times as he could. He had relieved his soul of countless years of suffering before my mother asked him to indulge her by sitting still when we went to see my father's sister. Without the distraction Michael created by getting up and going out, each visit seemed endless, and I began to understand more clearly the concept of time passed in Purgatory, though I still wondered why God had designed a system of punishment that would have absolutely no effect on us. Why would God send our souls to a fiery pit? Since our souls possessed no physical properties, there was nothing about them that would burn. When I asked my mother about this, she said it was one of the many mysteries God gave us. Because it increased our faith, whatever we didn't understand was a special gift to us from God.

One of the things that later became a mystery to my mother was why I started shoplifting in my sophomore year of high school. She couldn't figure out why, on my father's birthday that year, I decided to steal a pair of navy-blue knee socks with red-and-yellow toes. She might have understood, she said, if I had stolen something for my father, but the socks were for me.

"I thought I knew you," she said. She was sitting at the head of the kitchen table, in the chair she sat in when she paid the bills and made out the monthly budget. It was a different chair from the one she sat in to eat dinner. It was my father's chair.

I didn't say anything.

She peered at me over the half-glasses she wore when she wrote out the checks, carefully balancing the bills she had to pay right away against the ones she could put off a little longer. My mother managed money so well that even during periods when my father was unemployed, we never lacked anything essential. We even enjoyed what my mother called "extras," going out occasionally for dinner on combinations of fast-food coupons and specials, or to free shows she saw advertised in the paper. Once, she relentlessly sold church raffle tickets all over the city to secure the prize given to the person who sold the most: a family membership for the pool at the Howard Johnson's motel near our house. We could swim there any time we wanted, just as if we were paying guests. She had outsold the runner-up by more than a hundred tickets. She did it again the next summer and the one after that. She used all of her energy not only to make ends meet, but to give us things we would enjoy. It shamed and astonished her that I would steal.

I should have told her then that I didn't steal because we were short of money. I had a small career in shoplifting before I got caught, and every time I stole, I had enough money in my pocket to pay for what I took. Though the things I stole were usually practical, they were never things I needed. I should have told her that I didn't steal because of what we were missing. But I didn't know why I stole. I couldn't explain the urgency and desire, the sense of danger, that came over me. I couldn't tell her how guilty and confident it made me feel.

Instead, I stood before her, speechless, letting her assume it was my first offense. "I've lost faith in you," she finally said. That was my only punishment: hearing those words from my mother. She must have understood their strength. I never stole again. So it was a mystery to me why I suddenly wanted to steal, eight years later, during the weekend Kelly and I traveled together to Kentucky, the summer after Sean died.

In the car from St. Louis to Louisville, Kelly kept a list of the rivers we crossed.

"Why are you doing that?" I asked. She'd found a scrap of paper and a pen in the glove compartment and left them lying on her leg, ready to record the name of every river we encountered, any body of water, really, that formed the slightest stream. Creeks counted. Lakes did not.

"I don't know." She shrugged. "I thought it might be fun."

It was fun. We looked forward to finding water, and it passed the time we spent not talking.

"One's coming up," I'd say if I noticed a bridge ahead, and we'd start looking for signs. Most of the rivers were marked, but some required us to pull off the highway and drive upstream or down until we passed someone fishing or swimming or sitting by the bank. Kelly would roll down her window and inquire, and most often the people she asked would come closer, lean against the car. "This river?" they'd say, and they would tell us its name, supplying us, sometimes, with other facts: where it stopped and started, what rivers it ran into, how dangerous, how deep.

"It's a dead river," one man remarked.

"The Dead River?" Kelly began to write.

"No, it's a dead river. No bites," the man said. "Not even a nibble."

With a roster of ten rivers and a running count — six so far — of those with no names, we were late getting to Louisville. We stopped for lunch before driving south to the town of Loretto, adding the last river to our list. It was a Saturday, and we had been driving for more than seven hours, totally absorbed by the diversion Kelly had created. We could have driven happily across the country, I thought, never talking, never saying anything but the name of one river, then the next.

Kelly gazed out the window. She picked up the pen and put it down again. In our family, she was the one most eager to know the name of everyone to whom we were related. She could rattle off the whereabouts of uncles, aunts, and cousins far removed, as if it were only a matter of time before she'd meet them. And when she learned that someone distant had died, she'd matter-of-factly mourn the missed opportunity. "It's a shame," she'd say, "we won't get to meet them." Years later, as a new mother, she would extend this opportunity to her children, presenting them to as many relatives as she could muster. "It's a shame," she'd tell me when the oldest of our great-aunts died, "that I didn't get the baby over there to meet her." Her son was eight weeks old, our aunt near ninety. It was the confluence of these two lives that meant something to her. Traditions, continuities, family connections — all of these meant something to her. In her spare and unsentimental manner, they were hers to remember and protect.

We were late, and my aunt had started to worry. We carried

our suitcases to the second floor of the novitiate, and my aunt asked if we wanted to sleep separately or share a room.

"Share," Kelly said.

My aunt showed us to a room with two iron beds and left us to get settled.

"What's a novitiate?" Kelly asked while she unpacked her suitcase. I didn't unpack mine. We were only staying for the weekend.

"A place where novices live."

"I guess we're novices now," she said, bouncing on her bed. The box springs squeaked loudly.

"Be quiet," I said.

"Why?"

"It's a convent. You're supposed to be quiet."

"We can't talk?"

"Talk," I said. "Don't bounce."

"In case you haven't noticed," she said, continuing to bounce on the bed, "we're practically the only people here."

My aunt knocked on the door and took us to meet the other nuns staying at the motherhouse that summer. When she joined the convent, in the early 1950s, the novitiate was filled with novices preparing to take their vows. Now, only a few women joined the order each year, and the building echoed with emptiness. After walking us through the novitiate, my aunt took us on a tour of the infirmary, the retirement home, the chapel, and a small cluster of huts in the woods, where a group of contemplative nuns sequestered themselves to pray and meditate all day. When we passed the huts, I could feel the presence of the women all around me.

Kelly sensed it too. "What's going on here?" she said.

"Quiet," my aunt said, and she took us by a shortcut to the swimming pool.

"Wow!" Kelly said. "Nuns."

It was my thought too. Each of the women in the pool wore a plain black bathing suit and a white swim cap. It was as if the basic style of a nun's habit — long abandoned by my aunt and most of the other nuns we knew — had resurfaced in their swimming attire.

"Did you bring your suits?" my aunt asked.

"Sure," Kelly said.

My aunt looked at me.

"She doesn't swim," Kelly said.

We ate dinner in the cafeteria of the infirmary, and afterward Kelly went for a swim. I watched her from the second-floor balcony of the novitiate while my aunt played scales on the piano in the parlor. Before it grew dark, several nuns joined Kelly in the pool, and the sound of their laughter flowed upward, meeting the repetitive notes of my aunt's piano and creating a sound both familiar and frightening. It was the sound of who I was becoming, an evaporating joy on one side, impenetrable monotony on the other. I sat in the middle, the sounds on either side, inside and outside. It was not madness. Was it madness? It was the advance of darkness, the strands of laughter becoming thinner and thinner in the evening air. I closed my eyes against it. The liquid sound of my sister. Would I lose her?

Yes, for a while, a few years later, I would lose her. I would stop watching her. It must have seemed to her that we had all stopped watching, and desperate for our attention, angry and envious of Sean, she would remind us of her presence and her pain. Did it matter that what she did may have been more an act of drama than of depression, more spiteful than suicidal? It didn't. She took the same risk Sean had taken, the same overdose as his. The same drugs. It was the same season, even; it was New Year's Day. She was eighteen. I arrived at the hospital ahead of the ambulance, ahead of my parents, ahead of everyone else. Anticipating her arrival, the staff asked me to admit her. I'm not her mother, I told them. Any relative, they urged. So I signed her in. Later, in the critical care unit, I watched as the nurse helped her to sit up and sip some foamy blackness from a glass. Can you do this? the nurse asked me. Can you help her finish this? Her stomach had already been pumped. I held the glass to her lips. It was filled with liquid charcoal. It seemed a barbaric antidote to me, one that would turn her urine black for days, hastening the outward flow of toxins from her body, or collecting them in one place so she would throw up. Like the charcoal filters Sean dropped into his aquariums, it was a foreign substance that other foreign substances would cling to. The black fossil of life, it allowed the fish to breathe free. Nights, we would turn out the lights in the living room — my father, Sean, Kelly, and I — and with the room lit only by the light of the aquarium, we watched the fish weave through the water, the neons, the clowns, the glass cats. It was the quietest of times. I remember the sound of our breathing, the aquarium like a life support. Now Sean was gone and Kelly sat silent before me, drinking the black liquid almost as if she were destined to do it. If I tilted the glass too quickly, I would cause her to choke, the charcoal foaming back out of her mouth. It took forever for her to finish. She kept her eyes wide open and watched me. If she were a fish, with her mask of whitest flesh, her black lips and unflinching eyes, she would have been oddly beautiful. The most exotic angelfish, she would have been the one we had to buy.

When my aunt finished playing the piano, there was only the laughter left and the sound of the cicadas swelling in the evening air. In the pool, the brightness of Kelly's suit stood out against the dark suits of the women around her. I watched her bobbing up and down in the water and wondered whether the nuns in the pool knew about Sean. We hadn't seen my aunt since the funeral, and I was relieved that she hadn't tried to talk to us about him. I hoped we would get through the weekend without the strain of that kind of conversation. My stomach tightened. I looked at Kelly. No, she was safe. They wouldn't say his name while they were swimming, while everyone was laughing and having a good time. But why was I worried? She wouldn't *let* them say his name. For six months, she had been willing the world not to say it. A few nuns were nothing against Kelly's resolve.

The nuns began to disperse from the center of the pool, some of them leaving, some of them starting to swim laps, and Kelly suddenly looked lost. She climbed out and stood shivering below the balcony.

"They're the contemplatives," she whispered up to me.

"Really? What were they talking about?"

"Baseball."

"Baseball?"

"Yeah. The Cincinnati Reds. Louisville doesn't have a team. I need a towel," she said.

"I'll get her one." I turned around, startled. After the sound of the piano stopped, I had expected to see my aunt join the others in the pool. I had been bracing myself against seeing that much of her body — I knew it would make Kelly nervous too — and I hadn't heard her slip in behind me on the balcony. I had no idea how long she had been sitting there with me in the dark.

Kelly came up and sat on the chair next to mine with her knees up under her chin and the towel wrapped around her.

"Are you cold?" my aunt asked.

"Unh-unh," Kelly said, rocking back and forth.

"Are you hungry?" There had been little in the cafeteria that Kelly would eat.

"Unh-unh," she said.

"Tired?" my aunt asked.

"I'm fine," she said, springing her long legs out from the towel.

We sat in silence until the contemplatives came out, still in their suits. They were carrying a cake and champagne and singing "For She's a Jolly Good Fellow."

Kelly stood up. "What's going on?"

It was a surprise party for my aunt. She had just finished a master's degree in religion, and as part of her program, she had studied dance as a prayer form.

"What kind of prayers can you dance to?" Kelly asked after she'd eaten two pieces of cake, accompanied by equal servings of champagne.

All the nuns except my aunt laughed.

"I'm serious," Kelly said. "Fast ones? Slow ones? Let's see." She stood up to dance but fell back into the chair. "I feel sick," she said.

"Maybe we should go to bed," I told her.

The room we shared was large, not a cell but a spacious room with arched windows draped in heavy brocade. Thick oriental rugs lay on the wooden floor. Only the furniture — two iron beds, two small dressers — matched our idea of a convent.

"Maybe there used to be more beds in here," I said when Kelly commented that the room seemed awfully empty. We were lying in the dark.

"Did you ever want to be a nun?" she asked.

"Sure," I said. "Didn't you?"

"Not ever," Kelly said. "Never."

At home, we had never shared a room, and talking to each other in bed, just the two of us, was a new experience. Mary and I had shared a room for nearly twenty years, and talking at night was the way we knew each other.

"I used to write letters to Aunt Nancy to ask her how to be a nun," I admitted.

"Mary told me you used to write letters to the saints."

I laughed. "Yeah, I did."

"Well, you might have been a good nun," Kelly said. "You could have been a contemplative. You contemplate things, don't you?"

I wondered how to respond. It was a question not unlike one I had been asked a few days earlier. Are you contemplating suicide now? the woman on the phone asked. It was the Sunday before we went to Kentucky. I was alone in my apartment, eating supper at the kitchen table, and suddenly, just as if it were a cold coming on, I felt suicidal. Out the window, I watched as a woman on the street, three floors below, untied a kayak from the top of her car. Every Sunday evening in the summer, I watched this woman come home and untie her kayak. She lifted it over her head and carried it between the buildings to her back yard. Though she lived next door, she was a stranger to me. She was a woman who went kayaking every weekend in good weather. I was a woman who watched her come home. I need to kill myself, I thought. The woman had short dark hair with streaks of gray. The straps of her swimming suit showed through her shirt. I remember thinking, as I watched her, that I knew exactly who I was. I was no one. I was nothing more than someone who saw things, who watched things from a distance and remembered every detail. My life was merely an accumulation of details, descriptions of distance, detachments, all stored up like a deep surface inside me, useless, except that I could say where the sun was in the sky the first time I caught a fish with my father; or how Michael looked the night he had his first boy-girl party in the back yard, how he carried an empty tray into the kitchen on one hand, like a waiter, and said with a grin to my mother, "More pigs in the blanket, please," a sign that his party was a success; or how, as I turned the aisle in Kmart one day, I was surprised to see Sean buying baby clothes for Sarah, how he looked at each little dress and smiled at the rows of ruffles on a pair of padded pants. I turned around and left the store so he wouldn't see me, and as I walked to my car, I passed his bike, locked to a lamppost in the parking lot. The sight of it overwhelmed me — the sweetness of him, his private acts of unclehood. The next day, he gave Sarah three dresses, sitting on the floor with her to unwrap them, and then dressing her in each one to see how she looked, cooing at her, telling her how pretty she was, a fifteen-year-old boy, a four-month-old baby. Even the happiest of memories made me sad. Happiness, sadness, joy, grief. They were all the same. They had become more like observations than emotions, things I could see and recognize and describe from a distance, but never fully feel.

I needed to kill myself. The thought came to me quietly, as detached and incidental as all the other details. Every other time, it had come in a fury of tears and despair. But this time, the thought came to me calmly. Was this how it happened, then? Is this how one knew it was time? When it came unaccompanied by an onslaught of emotion? I was sitting by the window watching the woman when it came to me. The calm conclusion: I needed to kill myself.

I sat there, so still, for hours. Outside, it grew dark. I closed the kitchen curtains Mary had made for me when I moved in. "See-through? Why bother?" she asked when I showed her the fabric. Diaphanous was what I wanted, but I didn't tell her. "Does that word have something to do with the moon?" Sean asked when he stayed over one night and saw the light coming through the curtains. "Yes," I said, thinking it did. But it didn't. "It should," he would say if I could tell him now. I fingered the yellow lace trim Mary had sewn on the fabric, insisting that the kitchen needed some color. "Don't kill me," she said when she hung them. The frilly band of brightness was meant for my own good. "Too much white," she said when she assessed what I had done with the apartment, my first. "Every room looks the same." It does, I thought as I sat there, wondering what to do, how to do it. Should I go to the drugstore and get a bottle of painkillers to swallow? After a while, I got up and dialed the number. I had written it down long ago and put it in my wallet, folded in half behind a photograph of Sean and Kelly. Are you contemplating suicide now? the woman asked. Yes, I said. Can you give me your number? the woman chanced. When the phone rang, it was my aunt, asking us to come to Kentucky.

Now, a week later, I lay there, still alive, as if none of it had happened. "I'm going to sleep," I said. I waited in the dark for Kelly's protest — "You didn't answer my question!" — but it never came. I looked over and saw that her eyes were closed, her chest rising and falling softly beneath the sheet.

The next day my aunt suggested we visit a Shaker village. Kelly had just studied the Shakers in school, and my aunt made a point of mentioning this to the contemplatives when we met them after Sunday Mass. They nodded their heads and looked at Kelly approvingly. "Did you know they invented the circular saw?" she said.

My aunt was excited to see that a program of Shaker songs

and dances was scheduled to start when we arrived. During one of the dances, she began doing the steps on the side, and then, confident that she had mastered the basic movements, she made her way onto the main floor. The dancers seemed surprised by her presence. A few of the younger ones exchanged grins in a way that made me feel embarrassed for my aunt. She looked large and ungainly, yet she appeared completely unselfconscious. Intent only on her feet and the floor, she seemed unaware of anyone around her. It wasn't long before she closed her eyes, and as I watched her carrying out her steps, I realized she was no longer fully there. Her movements were awkward, but she looked serene, and though I could not rid myself of my own embarrassment, and could sense, too, that Kelly was embarrassed beside me, I saw for the first time my aunt's spirituality, revealed to us in a human, flailing form on the dance floor.

Afterward, we toured the houses, barns, and work buildings, and everywhere we went, my aunt encouraged Kelly to ask questions. Kelly began to walk at a slower pace, putting some distance between herself and my aunt's persistence. She stopped in each room and stared at the placards of text on the wall as if she were reading every word. After the tour guide stopped talking, Kelly raised her hand. "Is there a gift shop?" she asked.

In the gift shop, a Shaker song played in the background while we browsed. "'Tis a gift to be simple, 'tis a gift to be free, 'tis a gift to come down where you ought to be."

"I'm sick of that song," Kelly said as she tried out some of the Shaker chairs.

My aunt had returned to the meetinghouse, and through the window we could see her receiving instructions from one of the dancers. They had stepped out on the common ground between the buildings. Side by side they stood, one woman dressed as a Shaker, my aunt a modern-day nun. From the gift shop we watched them, teacher and student, lurching from leg to leg.

"Let's buy some things," I said to Kelly.

We picked out gifts for everyone in our family. Beeswax candles and candy. Small boxes made of wood. A souvenir shoehorn. Mints for my mother.

"Maybe some candy for the contemplatives," Kelly suggested. I looked at her. "I like them," she said.

We were the Shakers' biggest sale that day, buying a few hundred dollars' worth of merchandise in a matter of minutes. When my aunt walked in, the woman was ringing us up.

"Candy for the contemplatives," Kelly said, picking up one of the boxes of chocolate she had stacked on the counter.

When the woman told us the total, my aunt looked at me, alarmed. I avoided her gaze. I often spent money to make myself feel things — happiness, here — and I didn't want my aunt, fresh from her spirited performance, to see what was missing in me.

She went to look through a shelf of sheet music, and while I waited for my credit card to clear, Kelly wound up some music boxes on display near the door. By the time I joined her, she had all of them playing at once.

"They all play that song. Except this one." She picked up a porcelain heart that stood out among the plain wooden boxes and held it to her ear. "The theme from *Love Story*," she said, holding it to mine. She put the heart down and picked it up again. "It's pretty," she said.

"I love that song," my aunt said. She was suddenly standing behind us.

"The theme from *Love Story*?" Kelly asked, holding up the heart so my aunt could hear it.

"No," she said. "'Tis a gift to be simple."

As we walked to the car, my aunt announced that we were going to a church dinner and a country fair that night in the next town.

At the dinner, Kelly ate some applesauce and two pieces of white bread with butter before she ran off to the rides. For the first time all weekend, I was alone with my aunt. I looked past her to the four-piece polka band at the front of the gym, where a man in shorts and suspenders was playing an accordion. My aunt wanted to know why my mother didn't make Kelly eat more.

"What does my mother have to do with it?" I said.

I wondered whether my aunt regarded my mother's resignation over Kelly's eating habits as a singular shortcoming or whether she thought my mother was unfit in other ways. For some reason, her remark made me remember the day my mother returned home from the hospital after the illness that caused her to lose her memory. Michael and Mary and I went out to the car to meet her. My mother's sister was keeping Sean and Kelly, who were then six and three. My father walked my mother into the kitchen; she sat down and looked around and didn't say much. She was holding a white octopus, a craft project she had completed in the psychiatric ward. "I made this," she said, wrapping one of the legs around her wrist. We all stared at it, eight legs of braided yarn shooting out from a Styrofoam ball. "I didn't do a very good job," she said, "but I told the teacher I thought I had small children." And with her eyes she wondered where they were.

Kelly returned to the table to ask my aunt if she would be able to keep a quilt if we won it for her. She wasn't sure whether my aunt was allowed to own anything.

"You have to put one of these on a number before they spin the wheel," Kelly said when we reached the booth with the quilts. She handed me a two-dollar chip.

"Where'd you get these?" I asked.

"The contemplatives," she said, and she nodded toward the next booth, where two of the contemplatives were betting on a trip to the Bahamas.

Kelly's lucky number was six and mine was nine; she put a third chip on the number five. "Five kids in our family," she said as she placed it on the board. It was the number my father always bet, citing the same reason. The words had come out automatically. They had years of habit behind them, and they hung there between us now, true and untrue. She kept her hand on the chip. I knew what she was thinking; I was thinking it too. Throughout our lives, when anyone asked the size of our family, we would have to decide how to describe ourselves. Should we answer as we always had, or should we say four kids now instead of five? There was something awkward about either answer. That's why Kelly's hand was on the chip, not because she thought she should move it or because she wondered whether our luck with that number might have run out. No, it was something much simpler. It was hearing the words out loud for the first time and realizing that they no longer fit us.

The man at the wheel called, "Bets down," and I reached over and took Kelly's hand, leaving the chip on the number my father always favored. "Father of five," I had once heard him say when describing himself to someone he hadn't seen since childhood. It seemed to me like a clear and enviable identity. The wheel spun, and when it stopped on a number that wasn't one of ours, Kelly slumped a little beside me, then ran off to join the contemplatives, who were heading toward the rides.

The next day, on the way back to St. Louis, Kelly checked the list of rivers in reverse. We crossed the Salt River, the Ohio, the Blue, the Anderson, and Indian Creek. After the Erie Canal — not the *real* Erie Canal, Kelly contended — we crossed the Wabash, the Little Wabash, the Kaskaskia, and once more the Mississippi. As we drove, I thought of the coincidence of my aunt's phone call inviting us to Kentucky, how I had answered expecting to hear the voice of the woman from the suicide hotline. I was neither sad nor happy to hear the voice of someone more familiar, and when my aunt asked how I was, I answered, as I always do, that I was fine. And though my aunt was the catalyst, it was not her call that sustained me. Or the hotline worker's call when she called back. It was the call I made to Kelly shortly after. "See you Saturday," she said.

When we got home, we took out our souvenirs from the Shaker village, and after we had distributed everything — my father's shoehorn, my mother's mints — I reached into the bag for the heart-shaped music box and handed it to Kelly. She was surprised and delighted to see it. "The theme from *Love Story*," she said as she held it to each person's ear.

Later, when I was leaving, Kelly followed me to the car. "Here," she said, holding out the list of rivers. I took the piece of paper and put it in my pocket. Unlike the heart I stole for Kelly, it was a pure and simple gift.

SALT RIVER
OHIO RIVER
INDIAN CREEK
BLUE RIVER
ANDERSON RIVER
WABASH
ERIE CANAL
WABASH RIVER
LITTLE WABASH RIVER
KASKALASKA RIVER
MISSISSIPPI R

CHICAGO

A few weeks after Kelly and I returned from Kentucky, the blond-haired boy arrived on our block. My mother thought he was a gift from God. "Doesn't he look like Sean?" she would say as she watched him through the window. Every morning, he walked up one side of the street and down the other, picking up people's newspapers from wherever they'd landed and placing them on their front porches.

Though he was the same size and maybe the same age, the boy looked nothing like Sean. But he had blond hair, and to my mother every blond boy passed for Sean that summer. The boy who circled our street every morning — no one knew who he was or where he came from — reminded my mother of Sean in other ways as well. His daily routine brought to mind

Sean's job selling newspapers on Sunday, and like Sean, he had a quiet demeanor. But it was clear too that he was preoccupied, obsessed even, in ways Sean hadn't been. He appeared to be wholly absorbed in his self-appointed task, and his apparent satisfaction with repetition, the simple pleasure of picking a paper up and putting it back down, suggested that my mother might be right. In addition to thinking that he looked like Sean, she believed the boy to be retarded.

Whoever he was, his sudden appearance on our street the summer after Sean died was for my mother a kind of miracle, a sign that Sean was safe, one she found comfort in until the boy disappeared a few months later, on the day after I drove Sean's clothes to Chicago.

It was my mother's faith that sent me to Chicago. She had been sending money to an orphanage there — the Mercy Home for Boys — since Michael was born. Years later, when Michael was nearly a teenager, she began including a request that prayers be said for her special intention. She made this request at home as well, asking Michael, Mary, my father, and me to pray for this same intention, the specifics of which she would never make known. Why was it special? I wanted to know. Would we recognize it when we saw it? Would she tell us when it came?

I prayed for my mother's special intention, but nothing seemed to happen. After a few weeks, I asked Michael and Mary if they had been praying too.

"No," Michael said, "I don't have that much money," and he showed me the prayer cards my mother kept in her closet in a box marked PRIVATE MATTERS — KIDS KEEP OUT. It was the first I'd heard of the Mercy Home for Boys.

"They send her cards because she mails them money every month," he told me.

"She pays for prayers?"

"Sure," he said. "That's what adults do."

Michael had, by then, a habit of figuring the cost of everything, and he tended not to waste his time on what he decided wasn't worth it. His conclusion that our prayers could not compete with those my mother paid for became a precursor of his total renunciation of Catholicism, a decision he would present to us at the dinner table a few years later. "If you must know, I'd rather be a Muslim," he would say, and the statement would remain for me a symbol of the differences between us: Michael, then sixteen, willing to say who he was, who he wanted to be, and I always weighing my words, becoming quieter and quieter, unable, finally, to say anything for fear it would give me away.

The year after Michael showed me the prayer cards from my mother's box of private matters, the mystery of her special intention was made manifest to us as Sean. Now, Sean's life having ended, my mother wanted to give the boys at the Mercy Home something more than money. She wanted to send them Sean's clothes.

It was a Saturday in August when I finally made the trip. I drove to my parents' house early to pick up the clothes my mother had packed, and as I turned into the driveway, I saw them inside the garage: three giant trash bags, the kind my father filled with leaves in the fall, freshly cut grass in the spring and the summer. "Where you headed with that bag?" he would ask if he caught us taking one from the shelf where he hid them behind his tools. "Those are outdoor bags," he would say, and reminding us of their price and their proper use, he would watch while we refolded the black plastic bag and tucked it back into the box. "Don't use your father's bags," my mother would routinely tell us. "Who's been into these bags?" he would ask if he noticed the slightest evidence of interference.

My father's car was gone, and I imagined my mother telling him to put out the bags before he left for work that morning. I thought of him carrying them down one at a time from Sean's bedroom, just as he had brought down bags of my belongings when I went away to college — yes, he had allowed me to use the outdoor bags, granting me a special dispensation, he said — just as he had carried down the bags for Mary when she moved. Though this time, instead of helping to deliver them to another place, a new beginning, he had left them lined up in the garage, where I would see them and transfer them to my trunk without much trouble, bags he had no hope of reclaiming. "That one looks like it has some life left in it," he would tell us after we unpacked, and he would refold it, placing it in a box of bags that had already been used.

I pulled up next to the bags he had set out. They looked as if they might contain the bulging overgrowth of our back yard, but they held Sean's clothes, Sean's shoes and socks and T-shirts. If I opened them, I would find his underwear, his gym shorts, his jeans. Packed within them were the corduroy pants he had gotten for Christmas. I had taken him to the mall so he could exchange them for another color. He took the blue ones back in their box, told the salesgirl that they were a gift, that he'd rather get gray. The girl smiled. "I bet you look good in gray," she said.

He took the pants to the dressing room and came out wearing them, turning one way and then another in front of the mirror, checking the back, the front, the sides, the length, the waistline. He asked me how they looked. But how do they *really* look? he pressed when I said they looked fine. He tried them with his shoes on and with his shoes off. Are you sure? he asked again, checking himself once more in the mirror.

They were the first pair of pants he ever picked out for himself, and he acted as if his life depended on their looking good. How do they look when I walk? he asked, and I realized then that he was examining the pants against any possibility they held for ridicule, against all the anxieties of adolescence, as if his life were a long hallway crowded with kids who were waiting to comment on his clothing or his hair or his complexion, a long hallway he had to walk down every day, hurriedly, holding his breath. How could life be like that for him? I wondered. A boy like him, smart, handsome, athletic, the kind of boy who, when I was in high school, I always thought of as beyond reproach, someone who could walk down the hallway casually, unaware of the effect appearance had on everyone else's pace.

Watching Sean in the mirror, a decade after my own adolescence, I admit I felt a certain satisfaction in learning that he — and others like him, the whole group I considered elite — was not immune to the anxieties I had experienced, the fear I would be made fun of at any moment. But then I felt a certain sadness, not just for Sean, but at the knowledge that no one escaped.

That night at the mall, I saw myself mirrored in my brother, and I recognized the pain that was laid bare for him in buying a pair of pants. But I didn't see how serious the situation was for him. I didn't understand the extent of his sensitivity and self-loathing. I only thought to say, "You do look good in gray."

I got out of the car and untied one of the black bags, releasing the anomalous smell of clean clothes. "Those are outdoor bags," I could hear my father say, though I imagined that this time he had refrained from commenting when my mother took the bags inside, carried them upstairs, and set them on Sean's bed. She left them lying there for days before finally forcing herself to go through his drawers and his closet. She found the corduroys still with their tags and set them aside, thinking she'd return them.

When I had dropped by earlier in the week, I had seen them on the dining room table, in the place my mother put things that required action — things to be wrapped, mailed, mended, or returned to the store. I hadn't seen them since the night I took Sean shopping. Afterward, we went to Mary's to play Uno, a card game that Sean called "stupidly simple." He couldn't figure out why, at his age, he thought it was so much fun to play a game based solely on discarding cards that matched each other in number or color until you were down to your last card and obliged to yell out "Uno!" as a warning that you were one card away from winning. "I mean, when you think about it, it's pretty stupid," he'd say. But he loved the game, and so did Mary.

Sean brought the corduroys in so he could show them to Mary. He went into the bathroom and came out wearing them, turning his attention to Sarah after Mary agreed they looked good.

"See, Sarah," he said, sticking out his leg as Sarah crawled toward him. "See, Sarah, Sean's new pants. Corduroy," he said when she placed her hand on his ankle. "Corduroy," he repeated when he picked her up.

Mary's husband, Dan, came up from the basement. "Is this what I missed by not having sisters?" he said. "Fashion shows?" He took a bottle off the stove and dribbled some milk on his forearm. "Here. Feed her," he said, handing the bottle to Sean.

"Da-a-a-n," Mary said. She had a habit of saying his name this way, quietly and slowly. His quick, efficient energy was like a raw nerve in our family, one we would come to appreciate, but in the early years of their marriage, Mary would issue his name regularly like a soft reprimand. It was something we always expected to hear soon after he entered a room, and to me, his name began to sound as if it were Mary's way of introducing him to us again and again, getting us used to the idea of him, the first newcomer to our family, the first potential promise — the first potential threat — of another family growing up around us and away from us. "Da-a-a-n," Mary would say, and always, in response, he would shrug his shoulders, look around. "What?" he would say. "What? What?"

"Da-a-a-n," Sean said from the couch where he sat feeding Sarah. "Da-a-a-n," he said again, and then he finished with words my father often said to us for no reason whatsoever as he marched around the house: "If you want to be in this family, you have to fall in."

"Don't you?" Sean said, nodding at Sarah. "Don't you? Yes, you do. You do," he said, dipping his face down to hers and sing-songing these simple words over and over until Sarah stopped sucking and smiled.

That was the last night I saw him. When Sarah finished her bottle, he changed out of his new corduroys and stretched out on the floor with her, rubbing her stomach, rocking her gently on her blanket until she fell asleep. Mary picked her up and carried her to bed, and Sean lay with his knees up and his hands under his head. "Are we going to play or not?" he asked when Mary returned.

Sometimes, because I never saw him again, I mistake that night for the last night of his life, and I find myself feeling happy for him, relieved that his last night on earth was filled with the kind of simple contentment he felt with Sarah. I fool myself into thinking that his final hours were ordinary and uncomplicated, that they were pleasant and familiar, and most of all, that they were painless. And that's what I thought when I saw Sean's gray corduroys on the dining room table. I thought of his happiness that night, not of the way he had worried over his new pants earlier in the evening. I thought of his safety, his security, his tenderness with Sarah. I thought of him stretched out on the floor, swaying his knees back and forth as he stared at the ceiling, his face calm and blank. He put his knees down, propped himself on his elbow, and said to me, "I was just thinking. Sarah and I have the same doctor. Don't you think that's weird?"

"Why?" I said. The diagonal line of his body dissected the carpet into two triangles, his length almost filling Mary's living room. It was a small room, but there was something to be said for the fact that he nearly filled it. He was getting taller. In the six months since Sarah was born, he had grown so quickly that Michael had taken to calling him Uncle Stretch, and then, affectionately, just Stretch.

"I don't know," he said. "When I'm out with her, walking her or something, people sometimes think I'm her father. But when you think about us going to the same doctor, it's like we're just two kids in the same family. That doesn't seem right. I mean, I'm her uncle, after all," he said, lying back flat on the floor and setting his legs in motion once more.

"Well, he's a pediatrician, so you probably won't be going to him much longer," I said.

A week later his doctor — Sarah's doctor, Kelly's doctor, the doctor who had been Michael's and Mary's and mine — would come to his wake, walk directly to his casket, and stand

over his body for a moment before turning to find my parents'
faces. He was one of the few people, I noticed, who went without hesitation to view Sean's body, walking in with the same
authority as the priests from the parish, who went straight
to the coffin and made the sign of the cross on Sean's forehead before inviting the congregation in the funeral parlor to
join them in prayer. These men — the doctor, the priests —
brought a kind of focus into the room with them, a formality
and authority that made our presence there seem whole and
purposeful, reminding us of the physical and spiritual being
that was my brother. The doctor paid his respects to my parents, and on his way out he stopped briefly to look at Sarah,
asleep in Mary's arms.

That night — the night I think of as Sean's last — he left his
corduroys in my car. I noticed them when I was a few blocks
away, so I turned back, thinking he might want to wear
them when he returned to school after the Christmas break.
He was in the kitchen in his underwear when I walked in, his
upper body bare. Unlike Michael, who often roamed about
the house in his underwear, Sean was shy. Aside from the times
he went swimming, or on the hottest days of summer, when he
slept without a shirt, it was rare to see him other than fully
dressed. He was standing at the kitchen counter eating ice
cream from the carton, and I startled him.

"I'm not looking at you," I said.

I put his pants on the table, and he sat down quickly at the
other end. His skin was like my mother's; he was the only one
of us to tan. It was as if his skin was thicker than ours. Able to
absorb the sun's rays without burning or freckling first, it held
its color evenly and deeply, keeping him the slightest shade of
darker white all winter. It was flawless, the kind of skin that
had to be touched, and whenever I saw him without his shirt, I
had an urge to put my hands on his shoulders, to curve my
fingers around the muscles of his upper arms, to feel the perfect shape of him, the cool sensation of his skin. I could never
walk by him without touching him. I had a desire, whenever I

was near him, to be nearer. It was an attraction that started soon after his birth.

In the first few months of his life, Sean almost always cried himself to sleep. My mother maintained, as many mothers did at that time, that it was better to let babies cry a while than to pick them up and pamper them. This was supposed to make them stronger, less spoiled. One afternoon when Sean was crying, I went in and stood beside his crib. I had no intention of picking him up or trying to appease him; I didn't doubt my mother when she said it did babies good to cry themselves to sleep. I simply wanted to witness how it happened — the exact moment when he suddenly stopped crying and fell asleep. It was the same curiosity that made me lie down next to the moss roses in the morning hoping to see them open, or, failing that, which I always did, trying again in the evening to catch them closing back up. I leaned on Sean's crib and watched him cry, as if there were nothing human about him, and I noticed the soles of his feet for the first time, how perfect and unused they were. I put my hand around one foot, and it filled my fist, registering a gentle resistance. A few seconds later, he stopped crying and looked at me in a way that no one ever had, as if his eyes held a deep memory, the beginning of a bond between us. Often after that I would go into his room, close my hand around his foot, and hold it until he stopped crying. It was as if — at the age of eight — I had found the first key to my fulfillment, the desire to be looked at in that way, and from that desire the simple act, the mastery, of making something happen: the moss roses opening and closing, Sean falling asleep.

Because of my feelings for him, his perfect body, the pleasure of seeing and touching his shoulders, as I had once held his bare little feet, I am caught between guilt and gratitude by the haunting oddity that the last sight I would be granted of Sean was the sight of him without his shirt. It is as if death, any death, were a conflict made eternal, the everlasting reign of the Greek gods, the lesson we will always fail to learn, as if the supreme order of the world were only a means of providing each of us with a well-plotted way to wander: the mother's story, the son's story, the sister's story, the Sophoclean way the world works.

And so I am left with this story, the story of Sean exactly seven days before his death.

"Uno," he said when I put his pants on the kitchen table.

I took a Diet Pepsi from the refrigerator and sat down in the chair next to his.

"Uno?" he asked, tilting the carton of ice cream toward me.

"No thanks," I said.

"Uno?"

"Yes. I'm sure." I said. I picked up the newspaper to finish reading a story I had started at Mary's.

"Uno," Sean said, as if he were introducing a lengthy thought that began, "You know . . ."

"Stop!" I said. "I can't stand it."

"Uno?" He smiled.

"No, I can't." Earlier that night, Mary had taken the lead in the card game, yelling out "Uno" hand after hand. "Is that all you know how to say?" Sean teased. "It sounds like you're speaking Uno. Uno language." And he went on to speculate about whether people could communicate with just one word, and then, as was his way, he attempted to prove, to our annoyance, that they could. My coming back with his pants had given him the chance to continue.

"U-nohhh," he said, "u-no." He drew out the syllables, pleading with me to "Come on, just once."

"No," I said. "I'm sick of it."

He ate the last spoonful of ice cream and pushed the empty carton, sending it sliding to the center of the table. Not wanting to stand up wearing only his underwear, he wouldn't throw the carton in the trash until I left.

"You're trapped," I said.

"Uno." He crossed his arms on the table and lowered himself so that his chin was resting on his hand.

"Don't you have to sell papers in the morning?"

"U-no." He nodded.

"Shouldn't you be getting to bed then?"

He grinned up from his arm. "U-no," he said.

"You sound like the dog on that cartoon you used to watch."

"Uno-ooh?"

"Yeah, Scooby-Doo. You look like him too, shaking your head like that."

"Uno?" he said, crossing his eyes and letting his tongue hang out. He seldom made faces or risked the consequences of doing anything comical. Even laughing, he usually looked uneasy, unwilling to let go completely, as if he feared that someone would make fun of him for it. It had happened to him before. Some kid at school had laughed at him for laughing after everyone else had stopped, and he guarded himself against the possibility of being singled out again with the tight mask of taking a joke.

He sat up. "Uno-ooh!" he howled. "Uno-ooh!"

"Yes. Good Scooby-Doo," I said, patting his head.

"Uno-ooh! Uno-ooh!"

"Scooby-Doo, Scooby-Doo. Tell me, did you make a New Year's resolution to act stupid like this?" It was the second day of January. Two nights before, he had gone to his first New Year's Eve dance. In a week, we would find a letter he had left on his dresser that included these lines: "The holidays really stunk, especially New Year's Eve. At twelve I didn't even notice I was alone. The dance really was a drag. I acted like such a fool. I hate myself for even dreaming that anyone would like me or go out with me."

He grinned and nodded.

"This behavior is based on your New Year's resolution?"

"Uno," he said.

The clock in the family room struck one, and a bird came out and sang a single note.

"UNO!" he said, waving one finger in front of me. "UNO!" he said again, smiling, pleased that he was using the word to mean exactly what it meant, as if we had reached the juncture in his game where all the odds came even.

"Oh, good," I said. "Does this mean you'll quit?"

"I guess." He yawned widely, leaned back, and clasped his hands behind his head, as if his experiment with language had been exhausting, its conclusion an opportunity to release the little energy he had left.

"Are you going home soon? I'm tired," he said.

"So go to bed." I looked up and settled my eyes for too long on the hair that had just grown in under his arms. Without his usual "Stop it," he unclasped his hands and crossed his arms over his chest. "Really, Sean," I said, "it's nice of you, but you don't have to wait up with me. I'll be fine."

"You're mean to me," he said, reaching across the table for his pants.

"I know. I can't help it. You're so easy to be mean to."

He put the pants under his head like a pillow and closed his eyes.

"Corduroy, Sean. Corduroy," I said.

"Shut up," he said, smiling. "Shut up and please, please go home."

"Uno," I said. I threw the ice cream carton in the trash. "So what was your resolution?"

"None of your business," he said, half asleep.

"Oh, come on, tell me."

"No. Then it won't come true."

"It's a resolution, Sean, not a wish. You have to *make* it come true."

"I know," he said.

"So tell me."

He lifted his head; thin stripes from the corduroy crossed his cheek. "Uno uno," he said.

"God." I groaned. "Why can't you just tell me?"

"I just did," he said. "It's not my fault you're a failure at languages."

"If I guess, will you tell me?"

"Yes, but you'll never guess."

"I might. Say it again."

"Uno uno," he said, and he laid his head back down and shut his eyes.

"Don't fall asleep before I give up," I said.

"If I tell you, will you go home?"

"Uh-huh."

"Promise?"

"Yes."

He sat up and yawned. "My New Year's resolution is to be more fun. There. Now go home."

"Are you serious?"

"Yesss," he wailed. "You said you'd go, so go. I have to sell papers tomorrow."

"No, I mean is that really your resolution — to be more fun?"

"What's wrong with that?" he asked.

"Nothing." I took his hand and slid my fingers between his. "You're fun, Sean."

"No, I'm not," he said. I couldn't tell whether he felt as defeated as he sounded, or whether, because he was so tired, anything he said would have sounded the same way. "I *have* fun, but I don't start fun," he said. "Fun doesn't ever really start with me."

He was right. Spontaneity — fun — wasn't one of his strong points. I didn't say anything, and after too much time had passed in silence, he shrugged. "It's only a resolution," he said.

"Yeah. I've never kept one of mine for more than a week."

"This is my first," he admitted, as if he were revealing a first kiss or some other first act of adolescence; then he looked up at me and grinned, his eyes eager and trusting.

"Hey," I said, straightening my fingers so that our palms were pressed together, "given your rate of success at new ventures . . ."

"Yeah?" he said.

"When we're sitting here a year from now, I expect us to be having a much better time."

"Shut up," he said, and he smiled, pressing my hand to the table.

"Really," I said. "You have years to be dissatisfied with yourself. Look at me — I live in a semipermanent state of self-dissatisfaction."

"I don't want to grow up like you," he said.

"At least I'm fun." I rumpled his hair. Thick hair was the only feature we had in common. ("When you go to get your hair cut," he once asked me after coming home from the barber, "do they always say something about how thick your hair is? Like 'You have enough hair here for two people'? Are we supposed to pay them more money or something?" he asked with real concern.) I gathered his hair together at the top and twisted it so it stayed in place when I let go.

"Don't," he said, pulling my hand away. His shoulder blades rose and fell like embankments on either side of his spine, two small mountains above a riverbed of knotty vertebrae. I watched his back rising and falling, a rhythm that was almost as steady as sleep. His arms went out in two triangles, his corduroys in a bunch beneath his head. I looked at the rest of his body, the rare sight of him almost undressed, the perfect form of him, his underwear a close hug of cloth against his hips. It seemed he was asleep now. The whole house was quiet. Directly above us, my parents lay in their bed. We had heard them laughing earlier, the creak of the floor as one of them walked from the bed to the door and back again. We would find it locked if we went up. No noise from above us now that the sound of our parents had subsided into sleep. No sound for a long while except the two of us talking. I placed my hand on his back. He was beautiful. More beautiful, perhaps, than any of us would ever be.

"How can we tell her that he's dead?" my parents said. Mary would tell me this years after, her envy alive and undiminished. "How can we possibly tell her that he's dead?" they said, speaking of me, discussing it between them on the night Sean died, as Mary stripped the soiled sheets from his bed, feeling like a chambermaid, she said. It took her eight years to tell me, the words spewing forth over the phone, she in St. Louis, I living in New York by then. I didn't hear another word she said. *How can we tell her that he's dead?*

So it was obvious, the bond between us, visible in ways I had never realized. Some quality existed between Sean and me that made my parents single me out, their words settling into Mary, making her feel as if she didn't exist on the night Sean died, as if her loss were somehow less than mine. I listened to what I never knew about that night, the physical details of his dying, Mary selflessly wiping up what was left, his room covered with vomit.

"You weren't there. You don't know what it was like. It felt like shit," she said.

"Maybe they were only speaking logistically," I said, the line — *How can we tell her that he's dead?* — ringing over and over in my head.

"No," she said, "they were talking about you. While I was cleaning everything up, they were worrying about you. It was like I wasn't there. It was like I wasn't even related to him."

"But you *were* there," I said, not wanting to sound unsympathetic. "Don't you see? You were *there,* and I wasn't. You were home with them. I was twenty miles away, sleeping. They were talking about how to get the news to me, should they drive down or phone me in the middle of the night? I bet that's all it was. They were probably just speaking logistically."

"God. Why do you keep saying that? Why do you keep talking like that? 'Logistically.' Why do you keep saying that word?"

It was almost midnight. Lying on my bed, imagining what Sean's room must have looked like, I thought about the time I woke up in the middle of the night when I was six and threw up all over Mary. "Logistically?" I said.

"Yes," she said, more softly, and I listened to her snuffling up great gulps of air, as if she were trying to call back the tears she had just shed. Whenever I heard her cry, it took me back to our childhood, and I pictured her round-faced and freckled, her hair harnessed by both a ponytail and a headband, a style that on other girls might have looked indecisive but that on Mary signified an inarguable authority and sense of control — stubbornness, my mother called it. I would have these few moments now, while she finished crying and blew her nose, to feel equal to her again, two sisters talking on the phone, no long separations of life between us yet, all her sharp, sophisticated edges blurred, until she lit up a cigarette, inhaled deeply, and slowly exhaled, the sound of it renewing the distance between us, all the exponential ways in which she had advanced into adulthood, the two years between us becoming five, ten, twenty, husband, children, house.

I wondered why she had called that night and not another. Why now, eight years after? Had something happened that brought it back to her? I thought about the night Sean died. There were things about it that I had never shared either, things I would have told her had the night not ended with Sean's death. For me, it might have been better, less embarrassing at least, if my parents had arrived at another answer to the question that had bothered Mary so much, if they had phoned me instead of sending Michael. I remembered my mother telling me that when she came home from the hospital that night, all she wanted was to be alone, totally alone, to sit in her room by herself, with the door closed. "But I couldn't, of course. Sometimes, good or bad, we can't escape our circumstances." As soon as she said it, I understood that we were her circumstances, the rest of us. She had longed, for a little while, to be free of us, to be alone with the shock of it, the loss of Sean. I too wished I could have changed my circumstances that night, to have been alone when Michael arrived. It wasn't a clean night for either of us, I wanted to tell Mary. Was there any such thing as a pure experience in life, I wondered, one that occurred in isolation, untouched by the combustion of being human? I wanted to say something more to her, something meaningful. I wanted to let her know I understood. "I'm sorry," I finally managed. "I'm really sorry. I wish I had been there."

"You don't know what it's like," she said, "to feel left out of your own family."

"I'm sorry," I said again. "I really do wish I had been there."

"Is that all you can say?" she said. "You keep saying the same things. 'They're speaking logistically . . . I wish I had been there.'"

I didn't answer. I opened my eyes and tears fell down my face.

"Are you crying?" she asked.

"No," I said. It would have been too complicated to say why I was crying. I was crying because something always kept me from comforting her on nights like these, from saying the right thing, an inadequacy made worse by how good at it she was whenever I called her. She had all the right words. They rose from some wellspring of warmth and understanding within her, while in me that reservoir seemed always empty. Cold words like *logistical* rose up instead, words that were always loudly inappropriate. I offered her words of distance and detachment, all the wrong words, none of the words she needed. Mine were words as unfit for the occasion as "uno, uno," as if I had gone on talking that way for the rest of my life, enigmatically, never mastering the necessary inflection, the passion, the ardent desire to be understood.

I was crying because my sister's words — *Is that all you can say?* — reminded me of the last time I saw Sean, his words to Mary — *Is that all you know how to say? Uno?* — generating a game that would extend the time we spent together, keeping us talking longer that night, as if it somehow lent support to his theory that people could communicate with just one word. But could we communicate, could we have gone on communicating if we had only that one utterance to offer each other? Of course. Of course we could have. So what? So nothing, Sean would say.

I was crying because what Mary told me — *How can we tell her that he's dead?* — was like a missing link in my life, an answer to questions I had asked myself countless times since Sean died. Was our relationship the way I remembered it, or had I begun to make it up, fabricating an intensity between us? And if it was real, did my parents worry about it or appreciate it? Mary's story of that night, the words she repeated, still alive and agonizing for her so many years later, were like a long-awaited affirmation. I was crying because I wished I had been there that night. I was crying because I suddenly remembered Sean sleeping with me one night when he was little, his head pressed against my chest, the sweet heaviness of having him there. I thought he was asleep. His breath was warm and wet. I was about to pick him up and lay his head on the pillow. "If your heart stops beating," he asked me, burying his head into my breast, "will I be the first one to notice, or will you?"

"I'm not crying," I said when Mary asked again.

"Yes, you are."

"How did you happen to be there that night?" I asked. "I mean, did you just stop by or something?"

"No. Mom called and told me to meet them at the hospital, and I went back to the house with them after it was all over. After he died."

"Mom called Michael then too?"

"I guess so."

"You're lucky," I said.

"What?"

"You're lucky. You got to see him."

"I don't know that it was so lucky to see him. He didn't look too good."

"Well, you are lucky. Mom called you first. Doesn't that mean something? Of all the people she could have called after she called 911, she called you. She needed you. She wanted you there. She called you and she called Michael. She didn't call me. Doesn't that mean something to you?"

"I lived the closest."

"Well, you're lucky then, if that's all it amounts to. You're lucky you lived the closest."

"What's that have to do with luck? You chose to move all the way across town. I didn't."

"I don't know what we're talking about anymore," I said, to

keep her from saying what was obvious, that I had chosen since to move clear across the country. In situations of sickness and death, she would always have to assume the duties of the older daughter. She would do the dirty work — change the sheets, scrub the walls, remove the human residue. I would be the one who was summoned after, favored and fussed over each time I came home. I would do nothing. She would do everything. And always she would suffer the glory I gained by going away, just as I would always envy her ability to be there. And we would love each other like this for the rest of our lives, orbiting each other in enmity and admiration, our childhood like the center of the universe, a fixed point that could pull us back to a time when we were still together, when distance was nothing more than a mathematical equation, a simple problem someone could show us how to solve.

"It's the Hubble flow," I said.

"What?"

"Family life. It's like the Hubble flow."

"I don't know what you're talking about."

"The Hubble flow. It's when celestial bodies withdraw from each other because the universe is expanding."

There was silence, and in the silence I knew Mary was feeling the inferiority she always felt in school, where she had never done as well as the rest of us. I knew too that though I was thinking about the Hubble flow — why was I thinking about the Hubble flow? Was it because the moon was moving away from my window, stealthily creeping back toward the East River, as if it were retreating from something it didn't want to witness? Though I was, in fact, thinking about the Hubble flow, I knew that I had mentioned it to Mary not just to change the course of conversation but to make her feel bad, to make her feel stupid, to change the balance between us. I could do this at will, issue some inconsequential scrap of information that would make her feel inferior, that would register as fear and shame on her face and humble her so completely that, seeing it, I would have to tighten my throat to keep from crying. I could say something that would send her straight back to school, make her remember herself as a child walking with her head down in the hall, her average achievements rendering her somehow invisible on those days when we carried home our report cards and handed them to our parents. I could bring all this back to her with very little effort, slip into it without even thinking, just as she could do the same to me, saying things — "It just occurred to me how long it must take you to shave your legs," or "You'll probably find you go through makeup much faster than me" — that struck at the heart of my insecurities while accentuating the advantage she had in being attractive.

"I guess I don't understand," she finally said. "The universe isn't expanding, is it?"

"I don't know. I don't know why I said that. It was just something I was thinking about."

"Well, I'd *like* to understand it," she ventured, and her words made me cry uncontrollably, no hiding it any longer, the memory of that night and now her earnest desire to understand whatever I might decide to dangle in front of her, the raw pain I felt in wanting at once to reach out to her while at the same time belittling her, subtly, without a second thought.

"I know," I said. "I know you would. But it doesn't matter, it really doesn't matter."

"What's wrong then?" she asked. "Why are you crying?"

I didn't answer. I listened to Mary trying to comfort me, and I felt worse for it, for Mary's willingness to always be there. Before removing her nightgown after I threw up on her when we were kids, she had first helped me out of mine, and as we sat there on the bed, bare, I threw up on her again before my mother came in and took us to the bathroom. When we returned to bed that night — clean bodies, clean nightgowns, clean sheets — my mother placed a plastic basin between us, and I lay awake feeling this odd desire to do it again, to be sick again, while on the other side of the basin, Mary slept soundly.

Now, all these years later — a thousand miles between us — I listened to Mary comforting me. I wanted to say I was sorry, but it seemed so insufficient, so false somehow. In truth, I envied her the intimacy of what she had done the night Sean died. I had never cleaned up anyone's illness, and what did that say about me, about my ability to get close to people, so unlike Mary, whose list of such intimacies was long? That's what I should have risked saying — that I envied her. "What's one of your best memories?" I once overheard Kelly ask her. Kelly was on one of her campaigns, compiling an album of family information. "Washing sand out of Sarah's hair," Mary answered, and I smiled in the other room, envious, imagining what she must have meant.

Now she had called to confide in me, to tell me one of her worst memories — cleaning the walls the night Sean died, removing the physical horror of what had happened. And she had done it well. There was no sign of it, no smell.

I should have said something sympathetic, but I didn't. "Scutum Sobieski," I said instead, my tears subsiding now.

"You can see that from where you are?"

"I wouldn't know it if I could," I said, "and anyway, it's way past ten-thirty."

Mary laughed. "God. Do you ever forget anything?"

"How could anyone forget Scutum Sobieski?"

"Not him," Mary said, "the ten-thirty part."

"The ten-thirty part was my favorite part," I said. "Choose the biggest, brightest stars and constellations first. I don't remember the rest."

"The kids have the cards now," Mary said, referring to the set of constellation cards that Michael had when we were children. "Enjoy stargazing, one of the oldest human occupations," it said in gold letters on the box. "Choose the biggest, brightest stars and constellations first and get to know their unique patterns in the privacy and comfort of your own home. An original and easy way to know the heavens. For locating stars, cards work best if used at approximately 10:30 P.M. local time." We would turn out the lights in Michael's room and shine a flashlight on the cards, pierced with holes that marked the stars in each of the constellations. A simulation of the night sky shimmered across the bedroom wall while Michael,

Mary, and I, wearing our pajamas, made random attempts at naming the universe. We were always fresh from our baths. I associated the constellations with clean skin, the smell of soap, Michael's and Mary's knees touching mine as I sat tucked tightly between them. I hardly ever knew the names. "Scutum Sobieski," I would say when I was stumped, naming the small constellation of three stars that shone between Scorpio and Sagittarius, Mary's stars and mine.

"Scutum Sobieski," I said to Mary now. I used to say the words for no reason except to hear the sound of them, secret code words that meant something and nothing. I said them sometimes, shrugging, to say "I don't know." I said them to say thank you, you're welcome, goodnight — the oddest all-purpose words I knew, made no less mysterious even after Sean told me that Scutum Sobieski was the shield of the king of Poland, the three stars signifying the cross on his coat of arms. I didn't care. Scutum Sobieski would always be just a sound, an expression, something I liked to say, something — a trio of stars — between Mary and me.

Mary laughed. "I loved the ten-thirty part too. But why do you suppose they chose ten-thirty? Because it was kind of late but not too late?"

"No," I said, "it has something to do with the movement of the heavens in relation to the earth's hemispheres, so that if we had ever gone outside to actually *look* at the stars, that would have been the time they'd most closely match what was on the map."

"Oh," Mary said. "I always thought it was because it was after the news."

"You did? I always thought it was because that's when *The Tonight Show* started."

"You did not," she said.

"Yes, I did. I always thought *The Tonight Show* — whenever you heard that music — was the time the real night life started and so that was the best time to look at the stars."

"You're serious?"

"Uh-huh."

"I don't believe you."

"I swear," I said.

"Did I wake you up when I called?" she asked.

"No," I lied.

We were quiet for a moment.

"Mary," I said.

"Yeah?"

I wanted to say I was sorry. I wanted to tell her, in turn, what that night had been like for me. I had always wanted to tell her, but how could I do that now? How could I say that while she was wiping the walls of Sean's room, cleaning up the aftermath of his overdose, a man I had just met was making love to me for the first time in my life? How could I tell her what it was like to be awakened right after by Michael? I watched the man pick his clothes up off the floor and put them on. "Was he sick, your brother? Had he been sick?" he asked as he sat on the bed tying his shoes. "Yes," I told him, the words coming from my lips so easily it felt as if they had always lived there, like the truest of lies. "Yes, he was born with a hole in his heart. They never thought he'd live this long." "That's good," the man said. Michael waited in the front room while we dressed. "I mean, it's good it wasn't sudden, you know, that you could see it coming." How could I tell her what it felt like when he kissed me in the doorway and handed me his card? I closed the door. I looked at Michael. He looked away.

Had things been different, had Sean not died, I would have called her the next morning and told her — sister to sister — what had happened, that a man had made love to me, that I had made love to a man. Instead, sitting in my parents' house the morning after, I looked at her across the kitchen table and thought about the night we slept out together in the tent — the "dwelling" — we had made from the washing machine box the summer before Sean was born. She had agreed, for once, to be Frankie. Playing Beach Blanket Bingo, I longed to equal her Annette. "Oh, Annette, Annette," she was saying when I first felt it. I stopped the game to tell her. "Mary, do this, like this, hold your legs together like this and pull yourself in tight like you're trying not to pee." "Don't call me Mary," she said. "If you're going to be Annette, you have to call me — oh, oh, wow!" she said. "What is that? How did you?" And I said, "I don't know," and we did it again, lying side by side in the back yard, surrounded by the noises of a summer night, doing it again, not touching ourselves or each other, only the slightest, most secret movement, two sisters sleeping out together, a tent of our own making, the houses dark, the trees dark, the street lit by a single lamp lengths and lengths away from us. "It feels so good," she said. "It feels like love," I told her.

I wanted to ask her, now, how long it was before she'd done it again. Was it years later, or had she done it right away, a satisfaction so secret that even I, sharing a room with her until she was twenty, had never heard it? Or was she like me — forgetting the feeling altogether, discovering it again with someone else? And if she had forgotten, did she remember me when it happened, as I remembered her, lying beneath a man and wanting more and more of him but thinking more and more of her, lying in bed on a winter night, alive with this new sensation but transported back to the summer before Sean was born, to the last time we ever played Beach Blanket Bingo, the first and only time I didn't have to be the boy?

I wanted to tell her, but how could I tell her now? What could I say about the night Sean died, the guilt I felt, the close associations of sin and sex, death and desire? Say this, my best friend, Ellis, had told me as we stripped the sheets from my bed the night after. I was telling her what I would have told Mary — laughing, kissing, confiding every detail, lapsing back into life, as if Sean's death had not happened. It would be like that the weekend before he was buried, long moments when the focus of it would fall away, leaving ordinary life there, until the realization would surface again: Sean's dead. Ellis bundled the sheets into a ball. I showed her the man's card on my dresser. He's a lawyer, she said. She looked at me. She called me sweetheart. We lay down on the bare bed. Say this, she said when I confessed my feelings of guilt, of shame, that I was having sex — sinning, according to the way I was raised — while, across town, Sean was dying. Say this, Ellis said: for a Catholic it's the cruelest coincidence. That's all it is, darling. You Catholics, she said, kissing my cheek. She talked to me about practical things, about the cold snap that was coming, how we should wrap the pipes in my apartment before we left, how she'd stop by each day until after the funeral to make sure they didn't freeze. What else can I do; what else do you need? she asked, holding my hand as she lay beside me. Tell me the story of *your* first time, I said. We had our coats on against the cold, the covers and sheets in a hump on the floor. It was in her mother's bed, she said. The boy's eyes were blue; she had to show him what to do. Tell me about the day Sean was *born*, she tried. I don't know — I don't remember, I cried. It's okay, she said. It's okay; tell me anything. I'm cold, I said. She unbuttoned my coat, my shirt, unzipped her jacket, undoing one layer of us and then another. She opened everything, removed nothing, unfastened me with the efficiency of a rescue worker looking for a wound, until we were completely undone, our clothes opened like a cavity she would fill with her own skin. It's okay, she said as she lay on top of me, breast to breast, pulling the layers of cotton, wool, and leather in around us. It's okay, she whispered, until I fell asleep, the warmth of her body above me.

When I woke a few hours later — ten or eleven o'clock on the night after Sean died — the room was dark. How had she gotten up, without my noticing, to turn out the lights? Had I slept that soundly? I studied the stillness of her. "There's something about Ellis," Sean had said once. "When you're with her, you always feel safe." And it was true. She lived simply, straightforwardly, as if life required nothing more than mastering a series of survival skills. And she had mastered them. I marveled at the shallowness of her breath as she slept, her blond hair falling across my face, her body no bigger than Sean's. Her arms were stretched out above me as if she were trying to shield me from something, as if I were a child whose body was precious and small and in need of her protection. I thought of Sean's body, how embarrassed he would have been by the attention paid him the night before, some stranger's hands stripping him, another's trying to revive him, strangers begging him to breathe, his slightest response urging them on, until there was no response, until there was just his bare body lying on a bare table, abandoned, bruised, perhaps, by the force of their exertions, their desire to enter into him, to make contact with the mysteries beyond flesh and bone and muscle that make us breathe. When they failed to save him, I imagined them covering their efforts with a clean white sheet and calling in my parents to look at the whole of him one last time, their boy, his blood settling like a lifeless liquid within him. And did they uncover him then, my mother, my father, touch what warmth remained in him, touch him in places that I would have, his ankles, his toes, the flesh on the bottom of his feet? Did they touch his stomach, his shoulders, his chest? Or was it enough for them to look at his face, to touch his hair, to kiss his forehead? And Mary and Michael, did they embrace him? Did they lay their heads against him? Did they hold each of his hands?

I lay there thinking of his body, thinking of my body, feeling it for the first time beneath the body of someone sleeping. I lay there thinking of the night before, of Sean dying, his life belonging entirely to his body. And me, belonging entirely to my body that night too, a stranger's hands stripping me, entering me, touching me in places I had never been touched, my slightest response urging him on. Leave it to us, I said to myself as Ellis lay sleeping; leave it to Sean and me to be at the same place at the same time, letting it all out, living life that night as never before. The cruelest coincidence, Ellis called it. Yes, maybe, but it would always be there, the connection between sex and death, the loss of my brother and my virginity coinciding. Would I ever think of one without the other? I lay there, Ellis's body above me, thoughts of sex and death and sin coursing through me in the darkness like some long-lost catechism. Don't you know what it means to be a virgin? Mary asked as I tried to hold my own among a circle of her girlfriends sitting in the basement, every one of them, at twelve or thirteen, more sophisticated than I'd ever be. It means to have a baby without being married, I said. What, then? I asked, wanting at once to run from their laughter, afraid that if I did I'd never be cool, like Mary, like her friends. What, then? I asked. It's a woman who's really religious, Michael told me later. Where would we be? I wondered. Where would we be if it weren't for each other, Michael, Mary, and I, explaining the world, exchanging the facts, no matter the method — humiliation, deception, shame — learning it all little by little, learning it all nonetheless.

I lay there wondering what they were doing, my mother, my father, Michael, Mary, and Kelly. Why, having endured the duties of the day — the funeral arrangements and flowers — had I felt like fleeing, calling Ellis to pick me up after I had notified the neighbors? Had anyone felt the same way, wanting to leave, to run away, to return to the time before it happened? What were they doing now? Had my mother and father bought the Sunday paper, as they always did on Saturday night, parceling out their favorite parts? Was Kelly talking on the telephone? Had Mary gone home? Was she putting Sarah to bed? Would she lie down after with Dan, or would she fall asleep on the couch with her clothes on? Where was Michael? Had anyone thought of going to the movies, as I had, as I would all week, wanting, each day, to escape into darkness? Where were they all? Why had I left, first chance I could, not quite comprehending?

I lay there thinking of this, Ellis breathing above me, when the phone rang. I didn't answer. She stirred a bit. Aren't you . . . ? she asked. No, I said. Have you been sleeping? she asked. Are you cold still? Her mouth moved softly against my shoulder. Neither of us rose. Yes, I said (I've been sleeping). No, I said (I'm not cold). Too warm? she wondered, rising on one arm. I shook my head, closed my eyes. It's okay, she said as my cheeks began to glisten. She caught my tears on her fingertips. It's okay, she said. I'm so sorry, she said. I'm so sorry, sweetheart, she said, her words soothing me. It's okay. It's okay, she whispered. She kissed my ear as she spoke. She kissed my tears as they came, my eyes, my eyebrows, my cheeks. She kissed my head, my hair. It's okay, darling. It's okay, I'm here. I'm here, she said, kissing my ear as the phone began again, ringing another round. And as if I were rising to answer, I took her mouth into mine, I took her tongue. Ellis, whom I had always loved and never loved. I took her head in my hands, her hair in my hands; I took her mouth into mine, as the phone rang and rang, half ringing to a stop. It's okay, she said, kissing my neck. It's okay, darling, she said, as I removed the layers she had earlier undone, her jacket, her sweater, her shirt. It's okay, she whispered, as she lifted me lightly off the mattress, tended quietly to my coat, took off my clothes, the cold unnoticed now. It's okay, she said.

How could I tell Mary what that weekend was for me, making love to a man one night, making love to a woman the next? "You want a boy too," Mary had told me the night we slept out together before Sean was born. "Maybe," I said, knowing even then, when I was eight and she was ten, that desire for me was something less settled than it was for Mary. For her, desire would always have a definite edge, a describable shape, an acceptable sex: the opposite gender. I would be the only girl she'd ever kiss, and even then it wasn't me she was kissing but me as a male movie star, Frankie Avalon this night, Paul Anka, Rock Hudson, Elvis Presley the next. It was important to her that I be the boy, a specific boy, one she wanted to kiss; but most nights, when I kissed her in return, it mattered little to me whether she was my sister or Annette Funicello or Doris Day or Sandra Dee or even my favorite, Shelley Fabares. She was simply someone with whom I could fill in the ambiguous outlines of my own desire.

I wanted to tell her. I wanted to ask her if she had done anything that weekend while we waited for Sean's wake, anything ordinary, spontaneous or unexpected, anything like normal life. I wanted to say I was sorry for what she had gone through — I'm sorry, my father said, that you're all so young. I wanted to say all these things, but suddenly she wasn't there.

"What's your name?" a new voice said.

"Finkly Winkly," I answered.

"Unh-unh." Mary's younger daughter laughed.

"Uh-huh," I said. "What's your name?"

"Alson Edder."

"You're Allison Elder?"

"Uh-huh," she said. "I'm three."

"Unh-unh," I said.

"Uh-huh," she squealed.

"No," I said. "Allison Elder's sound asleep."

"Go seep," she said, and she hung up the phone.

I waited for Mary to call right back, and when she didn't, I imagined her falling asleep, Allison in her arms, and I wondered whether her memory of that night — Sean's soiled sheets, his walls, her feelings of exclusion — would be easier when she woke up. I envied her for having been able to tell me finally, an act of exorcism that might free her of those feelings — hatred, jealousy, rejection — that surrounded his death for her, while mine — ecstasy, embarrassment, distraction, desire, and, yes, a certain sense of shame — would linger, a sacrilege, something I would wear like a shield, protecting me from intimacy and honesty, keeping me quiet, alive but distant, like the stars in a city sky. I couldn't name that constellation, the emotional complexity of death and desire. I couldn't call it what it was: coincidence — circumstance — one night, consummation the next. It was nothing simple, like Scutum Sobieski, nothing one two three, like the symbols of Christianity that separated Mary's stars and mine. I envied her for everything and nothing, for Allison, *go seep, go seep,* for life and what's left in it. I envied her for having been there, for having stayed there, for having seen Sean's body on the night he died.

"I have to see him," I told my mother when Michael and I finally got to the house. "I have to go to the hospital and see him."

"You don't want to see him," she said, holding me in her arms. "You don't want to see him that way. He was so white. He was . . . he was just so white."

"No. I have to see him," I said. "Where do they have him? I need to see him before they do all that stuff to him. I have to, Mom."

"No," she said, stroking my hair, "you don't want to see him that way."

"Yes, I do," I said, pulling away from her. "Mary saw him. Michael saw him. Why can't I see him? I have to see him too."

"Just remember him the way he was the last time you saw him. That'll be better."

"No. I need to see him. Just once more before . . ."

"You can't," she said. "He's dead."

The last time I saw Sean he was falling asleep, his back completely bare below me.

"I'm going now," I said, kissing his head.

"Good," he said, without getting up.

I went to the opposite end of the table, put on my coat, picked up my keys.

"You're really going?" He raised his head from his corduroys. His right cheek was striped with sleep.

"Yes. In just moments you'll be free to roam the house in your underwear."

"I just want to go to bed."

"Yeah," I said, "and all along you've been lying on top of your ticket out of here."

"Huh?"

"Your corduroys."

"Ohhh!" he said, collapsing onto the gray fabric. "I never thought of that."

"Yeah. Pants," I said.

He shook out the corduroys and slipped them on without getting up from the table.

"You should've stayed in Boy Scouts, Sean. You'll never survive on your own."

"Go home," he said, and he grinned as he hoisted himself slightly off the chair, pulled the pants over his hips, and tugged at the zipper.

"Uno," I said, raising my hand goodbye.

"Uno," he said, standing up.

That was the last time I saw him, standing in the kitchen in his corduroys, and I remember turning around in the hallway, catching a final glimpse of him as he pulled at the waistband of his new pants, a modest boy who wished he was more fun. It was my last memory of him, one I returned to often, so it shocked me — eight months later — to see the corduroys on the dining room table, looking as new as the day he got them. It was as if they had been resurrected, part of the memory I held dear suddenly materializing before me, real and not real, a madness on my part, a mirage, as if that night, after I said goodbye, he had stepped out of the pants, left them folded neatly on the table, walked through the wall, and gone away.

My mother had retrieved the receipt from the archive of expenditures she keeps filed in the bottom drawer of the buffet. It protruded from the hip pocket, *Sean's Christmas* scrawled across the top. There were other receipts with *Sean's Christmas* filed in the bottom drawer, my Christmas, Mary's Christmas, Michael's Christmas, Kelly's, but rather than denoting my mother's simple system of purchases and returns, the words seemed sad and insufficient. What was left of Sean's Christmas? Not Sean himself but these corduroys, en route to becoming new currency in my mother's notebook, a line item in next month's budget for Kelly's school shoes. It was the way she survived, the way she sustained us. "Don't return them," I would say. "Don't be silly," she would answer.

At the end of the week I would drive Sean's clothes to Chicago, carrying them across state lines in the trunk of my car like some kind of contraband, the belongings of a dead boy, my brother. I would drive them out of Missouri and into Illinois; past the cornfields, the correctional facilities, and into the old coal country. I would drive his clothes through the heartland, through the wasteland, along the shoreline of the lake. And why would I do it? Why would I not refuse my mother's request, suggest that we dispose of his clothes closer to home? All summer I dreamt I passed Chicago and drove until I came to Canada, where I was told by the border guard to open the trunk. I hesitated and then complied. Each time I had the dream, I turned the key and Sean climbed out, brushed himself off, and walked away without me. Each time he was wearing clothes I didn't recognize. I had the dream four times that summer, and then, sleeping over in Chicago, his clothes finally delivered, I dreamt it one last time, and when I turned the key and opened the trunk, he climbed out completely naked, walked across the border, and turned around, offering me the full view of his body from a few feet away.

No one was home now. It was an afternoon in early August. The air conditioner was turned up high. It was cooler than cool, quieter than quiet, the corduroys almost cold when I carried them upstairs to bury them in one of the black plastic bags lined up beside Sean's bed.

"Thanks for saving me the trip to the store," my mother said when she called later. "I didn't know you were stopping by. I was out running errands. I would've waited."

I imagined my mother making a notation in her notebook that night. None of the dollar amounts was entered without explanation. "Sean's pants, returned Aug. 3," she would write next to the entry.

"I was surprised when I came home and found the money on the table with your note. Did they give you any trouble?" she asked.

"No. None."

"Well, they shouldn't have. Those pants were brand-new."

The following Saturday my mother came out to the garage in her housecoat as I loaded the bags into my car.

"You have a pretty day for a drive," she said.

I tried to close the trunk.

"Maybe you need to put one up front," she suggested.

"They'll fit," I said, laying one on its side.

"You're just like your father," she said when the trunk sprang open. After she went in, I rearranged the bags again. I didn't want one in the car where I could see it.

When I went into the house, I found my mother at the kitchen table wearing the half-glasses she wore when she made out the bills. I sat next to her and watched her add a long column of numbers. She wrote the sum on a scrap of paper and added the numbers again. Aside from praying and dusting, adding was the only act that totally absorbed my mother, and to watch her tally a long list — she never used a calculator — was to witness what she might have looked like had she stayed in the School of Commerce and become a bookkeeper, as she had planned to do before she married. As a child, whenever I saw her sitting at the kitchen table with the bills spread out around her, I knew I could sit right next to her and never be noticed. Instead of being bothered by her inattention, I felt calmed by it. Sitting beside her, unacknowledged, I felt as if I were in some special zone, a quiet, impenetrable place she had created with her concentration.

And so, the car already loaded, I sat beside her, happy to be enveloped in her disregard, the suspension of time that came with it. After she'd added the numbers a second time, she placed a check mark next to the total and peered at me over the top of her glasses. I smiled. Did she know how soothing she could be, how she could shrink the world down to this for me? Did she know that my memories of learning math were mingled with the smell of her, the breath of her, the way she whispered homework answers in my ear? Did she know that my attraction to things I could not comprehend came from her, from the sensuousness of her assistance? I would do anything for her. I would drive Sean's clothes across the country if it came to that.

"Here," she said.

"What's this?" I asked. "Boys' Clothing Donated to Mission of Our Lady of Mercy, Chicago, Illinois, in August 1982, by Lois and Tom Finneran," it said across the top.

"I need you to give that to them and ask for a receipt. I can deduct it from our taxes. It's a charitable donation."

My mother was asking of me the thing I most hated to do. I

had always been shy about these kinds of transactions, and easily embarrassed. If she knew this, she would have realized that I hadn't returned Sean's pants.

On the list was the denim jacket my mother had bought Sean for Christmas, the one that he really wanted. A few lines down, his new basketball shoes. They had barely been worn. There were two pairs of hiking boots. Five pairs of jeans. Five undershirts. Three sweaters. I put the list in my pocket, trying hard not to cry. What could I say to make my mother change her mind? What if I lied to her? I could say I forgot the list. I could lose it. And then it occurred to me: I could just forge a receipt.

As I was thinking this, my mother was fumbling through her address book looking for the Mercy Home for Boys. One of the few parts of her life that lacks order, her address book is an uncontainable catalogue of everyone she knows. A few pieces of paper fell out and fluttered to her feet.

"I don't live here anymore," I said.

"I know," she said, stuffing my old address back in the book. She pulled out a scrap of paper on which she had once written the address and phone number of the Mercy Home.

There was nothing now to keep me from leaving.

"Stop along the way to eat something," she said, pushing a ten-dollar bill in my palm.

I had my hand on the doorknob. My mother turned to me, her eyes full of tears. "Your brother had a beautiful body," she said, hugging mine.

As I backed out of the driveway, I hit the blond-haired boy. He appeared out of nowhere, fixed his eyes on mine in the rearview mirror, and with a shocked look on his face, fell backward. I leaped out of the car — God, please God — and found him curled up on the concrete laughing, slapping a newspaper against his leg. "Got you good, got you good," he said, his mouth contorting wildly around the words as if it took all his facial muscles to make them. I was relieved and annoyed at the same time.

After a few minutes, he stopped laughing, lay completely

still on the concrete, and closed his eyes. Up close, I could see clearly that beyond his blond hair and the size of his body, he looked nothing like Sean. As I stood there, the color began to fade from his face, and his breathing became slow and shallow. By the time my mother came out, he was white.

"You hit him?" she cried. "How did you hit him? Why weren't you looking?"

Had I hit him? Hadn't he been joking? The boy wasn't moving, and the speed with which his face had lost its color seemed abnormal now. Had I misread everything? Maybe he hadn't been laughing. "Got you good, got you good." Had he meant that I had gotten *him* good? I thought of the way Sarah, only a few weeks earlier, had started calling herself "you" after we'd pointed at her in photographs and said, "That's you." Could the boy have the same level of language as a child just learning to speak?

"I don't know," I said. "I thought he . . . I thought . . . I don't know."

My mother bent down. "Are you okay, son?" she asked, and just as she put her hand on his shoulder, I saw a tremor along his lips, a tension in his jaw, as if he were trying to hold back a smile.

"He's playing dead," I said.

The boy broke into a full smile and opened his eyes, but the rest of his body remained unresponsive.

My mother straightened up. She was wearing a cotton housecoat, and her feet were bare.

"What's your name, son?" she asked. "Do you live near here?"

The boy kept smiling. He was staring away from us, his eyes fixed on our front porch. On either side of it, one of the concrete statues my father had bought for my mother's birthday that month peeked out, like a shy child, from below the bushes. My father said that one statue reminded him of Kelly, the other of Sean.

My mother looked at me. What's the matter with him? her face asked.

"Can you get up?" she said to the boy. "Are you able to walk?" And as if he'd been waiting all along for some direction, he reached out his hand and wrapped it around my mother's ankle.

"Got you good," the boy repeated, and he strained to get up, sliding his hand up my mother's leg to steady himself. Beneath her housecoat, my mother was naked, and as I watched the boy move his hand up her leg, I remembered what it was like to be a small child, the smell of my mother in the morning, my head not yet reaching her waist.

"Heh," he said when he stood up. "Heh." It was a pure and primitive utterance, a physical vibration that showed in his throat. "Heh," he said a third time, and he handed my mother the newspaper and walked up the hill to the next house.

"You better get going," my mother said, and without looking at me, she went back into the garage and pushed the button that released the garage door, sending it creaking and curving over its tracks. I watched her body disappear, segment by segment, as the door unfurled itself between us. When it reached the concrete, separating us completely, I got into the car.

In a truck stop halfway between St. Louis and Chicago, I watched a woman steal two candy bars and a tube of toothpaste before I went into the diner and ate a piece of coconut cream pie. "How is that, honey?" the waitress asked when she came back a second time, ready to refill the coffee I hadn't ordered. As I ate, I thought about the shoplifter. Candy bars and toothpaste. There was a kind of symbiotic relationship between the things she stole. Had she ever gotten caught? I wondered. Did she have a mother who had lost faith in her? I remembered the shame my mother felt when I shoplifted. Had I ever told her I was sorry? Was I sorry? What did it matter anymore? I had shoplifted again. I had shoplifted from the Shakers. I was the lowest form of shoplifter, stealing from simple, spiritual, unsuspicious people. I had shoplifted again after that, at Kmart the week after Kelly and I came back from Kentucky.

"You're tampering with your chances," my mother might have told me had she known. Yes, by shoplifting — by sinning — I was tampering with my chances, taking risks she herself would never have taken. A few months after Sean died, she lost her job. She had been working part-time in a doctor's office, keeping his medical records. When the doctor decided to computerize the office, he offered to send her to school to learn the new system, but the idea made her nervous. "I'm too old," she told me.

She asked the doctor for another position, one that wouldn't be affected by technological advances, something safe and small. She liked the man. She needed the money. He was sorry, he said. He had nothing else to offer. On her last day, he told her he had no intention of interfering with her eligibility to collect unemployment, and he urged her to apply.

She thanked him and said she understood. Weeks went by, but she never applied. The doctor called to ask how she was doing. Did she have another job? He'd never been contacted about her unemployment, he told her, and was worried that she may have misunderstood. He was calling to reassure her: he hadn't planned to contest any claim.

She thanked him again, but still she didn't apply. Instead, she pored over her budget, looking for ways to cut corners, to manage with less money. "It wouldn't be right for me to get unemployment," she told us. "I didn't get fired. I wasn't laid off. I was offered a new opportunity, but I was afraid. You're not entitled to collect if you *choose* to leave your job."

"Don't you understand?" we explained. "The doctor's going to say your job was eliminated, that the workload lessened and he couldn't use you any longer. He's going to tell them you had no choice in the matter."

But still she didn't do it, and while she looked for another job, the stress of having less money grew greater every month.

"You don't get it," we told her. "There's no reason to put yourself through this. The doctor's fixed it so you can collect."

"*You* don't get it!" she yelled back, crying. "I have to follow the rules. I have to do what's right. I can't take any chances.

Don't you understand? I have to get into heaven. I have to see him again!"

While my mother was hedging her bets for heaven, I was tampering with my chances. What would she make of it? What would she say if she knew who I was, what I was doing?

I ordered a second piece of pie, and the waitress eyed my body as she wrote down the order. It was not yet noon. I was thinking about what I would do after I'd dropped off Sean's clothes at the orphanage — whether I would visit some of my favorite places in Chicago or venture out into some neighborhood that night that I didn't quite know. Maybe I would just hang out in my hotel room, order room service, and watch twenty-four hours of movies on TV until it was tomorrow, daylight, and I could drive straight back to St. Louis, mission accomplished. Maybe I would walk along the lake late at night, or go up to Northwestern, my sort of alma mater, sit in the student union, and see whether I had buried the failure of not having finished. Maybe I would ride the el from one end of town to the other.

I thought about what Sean would want to see in Chicago if he were with me. We would probably go to the aquarium. After looking at every tank, we would show each other our favorite fish. As always, Sean's would be some unsightly specimen that camouflaged itself so well it took a certain skill to see it. I would give up, and he would put his hands over my ears and move my head until he swore I was staring straight at it. "It's right in front of your face," he would say. I would shake my head, laughing. Some kid would squeeze between us and agree with me that there was nothing there. Then Sean would ping his fingers on the glass a few times, trying to coax the phantom creature out, knowing full well that the sound would make the fish retreat farther into its hiding place. But there would be movement, some stealth that caused the sand to shift slightly, or the mud or the mass of seaweed, giving us reason to watch a little longer, the promise of something present that might actually appear. Sean would tap again in front of the fish, then tap a third time someplace else, frightening the fish into moving forward. "See," he would say as it swam past us to its next concealment. That's how it always was with Sean. To spend time with him, you had to be patient. You had to be willing to wait.

"Sit up straight, son," the man in the booth in front of me said, pulling up his son by his armpits so the child sat taller at the table.

It was what my mother had called the blond-haired boy that morning. "Son," she had said. I had never heard her address anyone as "son" — certainly not Sean or Michael. My parents have never referred to us, or to each other, by anything but our given names. Their voices, the way they talked to us, the things they sometimes said, made clear their love for us, but we never heard in our house the words *honey* or *sweetheart* or *darling* or *dear*. My father called me Stinky when I was a child, and once I heard him laugh and call Mary Sunshine. She was red all over from lying out too long at the pool.

So my mother's use of *son* that day was surprising. She had spoken it warmly and with concern. "Son," she had said, as if she were testing some new territory, trying to reach out to the boy whose blond hair reminded her of Sean. What else could she have called him? No one knew his name.

"What else can I get you, honey?" the waitress asked as she stood over me, writing out the check. I looked up. A line was forming for lunch.

Between the truck stop and Chicago were a few more farms, a prison, the coal towns, and the suburbs. When I reached the city, I checked in to a Holiday Inn near the lake and asked the clerk if he had a room with a view.

"Of what?" he said.

"Of the lake."

"Do you have a reservation?" he asked.

"No."

"You can't get a view unless you've reserved a view," he said.

"Can I get a *room* without a reservation?" My voice sounded anxious, and the clerk looked up for the first time. He smiled, and, mistaking the smile for kindness, I let myself feel relieved.

"Of course," he said, "but views are for people who plan ahead."

I could feel my throat growing tight. I kept my head down, and after what seemed an endless transaction, I took the key he set on the counter. Once inside the elevator, I let my throat loose, and tears streamed down my face. Why had I done it? Why had I allowed myself to ask for a view, made myself vulnerable? I tried to recall my brief satisfaction at summoning up the courage, saying the words. It was as if there were a stranger inside me, someone stronger and bolder than the person I always presented. I felt a certain power thinking that the hard part was over, that the only impediment to getting what I wanted was breaking through the fear that prevented me from voicing my request.

The room was warm. I pulled back the curtains to open the window. It was sealed shut and faced a brick wall. "Buff," my father would say if he saw it. "Hard brick to sell in St. Louis." "Buff," I knew he would say even if he were alone.

I lay down on the bed, feeling belittled by the encounter with the desk clerk, the ugly emptiness of the room. I thought of the many times, as a child, I had witnessed my mother being mistreated in similar situations, patronized and reprimanded. There was the man who scolded her when she picked up a piglet in the petting zoo and carried it away from a crowd of children who were pawing at the other piglets in the center of the pen. She had wanted to show Mary, who was cowering in the corner, how gentle and harmless piglets were. Seemingly taking it in stride as we toured the rest of the tiny zoo, she cried quietly in the car as she drove us home, sinking further and further into herself, until one of us brought her back; this time it was Michael, oinking beside her in the front seat, louder and louder until he made her laugh. Or the woman who said, "I doubt that *that's* a good idea," when my mother approached her at the pool and asked if we could try out the giant inner tube that was floating idly at the woman's side. Or the store clerk who wouldn't listen to my mother's reasons for wanting to return a shirt she'd bought for Sean. She could tell he didn't like it. The woman kept pointing to a sign about marked-down merchandise that read, NO RETURN. NO REFUNDS. NO EXCHANGES. When my mother turned to leave, the woman rolled her eyes at her co-worker, as if to suggest that my mother couldn't comprehend the concept of a final sale. There was the nun who had rejected her evening dresses at the clothing drive. There was one person after another, one stranger and then the next. At an early age, I began holding my breath over each encounter, hoping that no one would hurt her.

My mother had been willing, again and again, to act in our interest, despite the pain it sometimes caused her. Was that the fundamental fact of motherhood? Why was it missing in me? Unlike my mother, who had longed for children, I knew that I would never have the true gift of life to give. I would be afraid for my children. I would be unable to act on their behalf.

I lay on the bed at the Holiday Inn for a long while, looking at the brick wall, thinking about children, wondering whether there was one of us my parents would have preferred to lose instead of Sean. Years later, I would find my mother crying in the living room. It was a Sunday. She had just come home from church. What's wrong? I asked, and she told me about listening to the story of Abraham and Isaac at Mass that morning. For the first time, she said, she recognized that her faith was equal to Abraham's; she knew that if God asked her to give up one of her children, she would come just as close as Abraham to obeying God's command. Was it the horror of thinking she could kill one of her children that made her cry? No, it was the horror of having to choose one.

"Abraham only had one child," she cried. "I have five. If I had to choose, we all know it wouldn't have been Sean." It was the truest thing my mother ever told me.

I fell asleep, and when I woke up I didn't know where I was. The sweat on the back of my neck made my hair wet and cool at the roots, and touching it reminded me of the sensations I felt as a small child waking from a summer nap, lying alone, listening to the sounds in my room and in the house and in the world outside my window. Now, I lay listening to the hotel room and the sounds from the hallway until I was sure, as I had been when I was a child, that no one would be coming to tell me it was time, that if I got up and opened the door, I would hear that everything was happening far away from me and that it was up to me to walk back into the world, to groggily reenter, still soft from sleep, and find the rest of my family — my mother, my father, Michael, and Mary — and by watching whatever they were doing, slowly become awake.

I got up and went into the hall. A maid's cart stood unattended. I stuffed my pocket with miniature bars of Ivory soap.

"Need anything else?" the maid asked when she came around the corner.

"No," I said, and I went back to my room, embarrassed.

It was late in the afternoon when I left. West Jackson Boulevard was easy to find, and in less than half an hour I was sitting in my car across the street from the boys' home, trying to get up the courage to go in. I looked at the list of Sean's clothes. The corduroys, of course, were not on it; as far as my mother knew, I had returned them. I thought about what else was missing: his favorite pair of dress pants, his sports jacket, his good shirt, the only tie he owned. He had been buried in the clothes my parents bought him six months earlier, for his grade school graduation. Looking at the list — at what was missing — made me think of the clothes Mary bought me to wear to his funeral. "Let's go shopping," she said. She knew without my saying so that I had nothing right to wear. I had no good clothes. I didn't own a dress. She bought me a skirt, a blouse, a jacket, and shoes. The jacket was part cashmere, the blouse pure silk. "You look so great," she told me when we went to his wake. She was kind to me that way all weekend.

I got out of the car and pulled a green football jersey out of one of the bags. The number 75 was printed in white on the front and the back. In the months after I bought it for him, he had worn it almost every day. I dropped it on the front seat. It was something of his I wanted to keep.

"You should've called to let us know you were coming," the nun at the front desk told me. My mother had called, but I didn't bother saying so. I set down the first bag and went out for the other two. When I came back, small piles of Sean's clothes covered the desk.

I let my fingers rest inside the waistband of a pair of his underpants. The piles of his clothing reminded me of the stacks of clean laundry we'd leave on one another's beds. I had often wondered how that routine got started. Was it my mother's way of respecting our privacy, not opening our dresser drawers? Or was it her way of saying, "This is as far as I go. I washed them. You put them away." Whatever its origin, I liked the casual intimacy it allowed. I liked the little-girl look of Kelly's underpants. I liked it when, in his twenties, Michael made a mistake and bought a style of underwear he wasn't used to wearing, cotton-knit shorts that fit tightly and covered his thighs. "I guess you noticed my new underwear," he said with a grin when I brought a stack of clean clothes into his room. I liked the thought of my brothers, whenever they folded the wash, placing my bras matter-of-factly on my bed. Most of all, I liked it that they knew without asking which were mine and which were Mary's.

"These look usable," the nun concluded.

"A lot of these things are brand-new," I said, as if I were delivering a message from my mother. I turned Sean's basketball shoes over to show her the soles, the treads completely uneroded. "And these," I said, smoothing my hand over his gray corduroys, "these I know were never worn."

"Most boys are harder on their clothes," she said, folding one of the flannel shirts she'd inspected.

"Yes," I said. "Well, he wasn't. He wasn't really . . . he wasn't ever really hard on anything."

"Oh, dear. He's dead then?" she asked. "I wasn't sure."

I nodded.

"I'm so sorry, dear," she said. She took my hand and held it. "He was your brother?"

"Yes," I said. I couldn't look at her without crying. I kept my eyes on his corduroys.

"Did he suffer?" she asked. "Was he sick?"

"He killed himself," I said, and I began to cry. I had never said the words so succinctly, and having said them, I couldn't stop crying.

"Let's sit in the chapel," she said.

I was ready to cry in the arms of a stranger, in the arms of anyone; I didn't care. She held my hand as we walked down the empty hallway.

I wanted her to stay with me. I wanted to sit for hours with someone in total silence. I wanted to fall asleep somewhere. "You stay in here as long as you like," she said, leaving me alone in the unlit chapel.

When I went back to the desk, Sean's clothes were gone. There was no sign of him, no sign that I had come in carrying anything but the keys I kept turning over in my pocket, no chance to look at his clothes one last time. No chance, again, to say goodbye.

"Feeling better?" the nun asked.

"Yes," I said. "Thanks."

"It's hard, losing a loved one," she said.

I nodded. "I need to ask you for a receipt," I said, and I handed her my mother's list.

She glanced at it nervously.

"Just for the estimated value of the clothes," I said, pointing to the check mark at the end of the list, where my mother had written the total. "For taxes. That's all."

"We don't ordinarily give receipts for something like this," she said. "For cash donations, yes, but not for clothing."

"It doesn't have to be anything official," I said. "If you just write out the value of the donation on a piece of your letterhead and date it, I think that would be fine." The words sounded resolute, but inside I was shaking.

"I'm sorry," the nun said. "If she insists on having a receipt, tell her she'll have to contact the director," and she began to write the priest's name on a piece of paper.

"She doesn't insist on anything," I said, and I took the list and left, letting my last words be this simple and accurate assessment of my mother.

Boys' Clothing Donated to Mission of Our Lady of Mercy, Chicago, Illinois in August 1982 by Lois and Tom Finneran

1 Denim Jean Jacket (NEW)	30.00
1 Nylon Parka with Hood (NEW)	65.00
1 pr. Converse Basketball Hi Top Shoes (NEW)	35.00
1 Corduroy fleecelined jacket	20.00
1 Rain Poncho	5.00
1 Hooded Jersey Jacket	5.00
1 pr. Spaulding tennis shoes	5.00
1 pr. buckskin sport shoes	8.00
2 pr. hiking boots @ 10.00 each	20.00
5 long sleeve flannel shirts @ 8.00 each	40.00
2 long sleeve cotton shirts @ 5.00 each	10.00
2 short sleeve cotton shirts @ 4.00 ea.	8.00
2 short sleeve cotton shirts @ 3.00 ea.	6.00
6 pair athletic shorts @ 3.00 ea	18.00
3 belts @ 1.00 each	3.00
2 pr. sweatsox @ 75¢	1.50
5 pr. blue jeans @ 6.00 ea.	30.00
2 pr. dress pants @ 6.00 ea.	12.00
5 undershirts @ .75 each	3.75
6 underpants @ .75 each	4.50
3 long sleeve sweaters @ 8.00 ea.	24.00
11 short sleeve Tee shirts @ 3.00	33.00
10 long or elbow sleeve Tee shirts @ 5.00 ea.	50.00
Estimated Total Value	436.75 ✓

I drove back to the Holiday Inn, slept for several hours, and then ate dinner in a coffee shop that was about to close. "Grill's already down," the waitress said. "I can only give you something cold." There were no other customers. The manager was counting the money in the cash register. "How 'bout tuna salad?" the waitress suggested.

"Sure," I said.

Afterward, I went to the movies. *An Officer and a Gentleman* had just opened. "Stupid fuck," a man behind me said when the guy who wasn't Richard Gere hanged himself. "Stupid fuck," he said a second time when Richard Gere held his friend's body in his arms, rocking him, kissing his head. As I watched, I wondered what it would feel like to hold a body that had no life left in it, and it occurred to me that when hugging the dead, you would always be the first to let go. And the last.

When I returned to the hotel, it was still early, so I went to the pool. It was open, but no one was in it; even the pool guard was gone. I rolled up my jeans and sat down, dangling my feet in the water. It was a new kind of loneliness I felt that night. I wondered what it would be like to go to the hotel bar. I had been picked up only once in my life. It was the night Sean died. I had stopped at a Dunkin' Donuts as I was driving home from the typesetting company where I worked. It was nearly midnight. I went in and bought a dozen doughnuts. I was going to eat them all myself when I got home. I wondered why I was doing this. I hadn't eaten anything excessive in almost a year. I was the thinnest I'd ever been. I was wearing my white corduroys and a maroon jacket with a fake fur collar that I had bought the day before. I knew I looked good, and I was feeling what I thought it must feel like to be sexy, so why would I stop at a doughnut shop?

A man sitting at the counter got up and paid for my doughnuts, and as soon as he did, I was embarrassed that I had gotten so many. He asked if I'd like to join him for coffee, and though I wanted to, I said no and went out to my car. There was still snow on the ground from the last snowfall, and a new snowstorm was just starting. The man hurried out after me.

"Why don't you come to my house?" he blurted out as I opened my car door.

"I can't," I said.

"Then let me come to yours."

"I have to work in the morning," I lied, and looking into his face for the first time, I was surprised to see that he was handsome.

I had always imagined that something like this would happen more slowly — and never at a doughnut shop. He closed my car door. "You're so beautiful, your eyes and your hair," he said, and he kissed me. "If you can't come home with me, sit in my car for a while," he said, and he pointed to the car next to mine. I was afraid, but I wanted to trust him. This may never happen to me again, I thought. This may be my only chance. I wanted, desperately, to be desired. Even more, I wanted to experience what it was like to be with a man, to discover how it felt to be what the world called normal: to be sexually certain, completely straight. "Let's just talk," he said, opening his car door. I got in. "I'm from Iowa," he said, and he showed me his driver's license. "I came down here to go to law school and I stayed. I have a second-floor flat over by the brewery." He rubbed his gloved hands together. It was cold. We could see our breath when we talked, but he didn't start the car to turn on the heat, and I took that as a sign that I was safe.

"You going to eat those?" he asked, pointing to the doughnuts.

"Oh, they're not for me," I said. "I was going to take them to work tomorrow. You want one?"

"Maybe in the morning," he said, and he kissed me again, deeply this time. I had never been kissed by a man, and I was surprised that I could kiss him so well in return. I could tell I was good at it, as if I had been kissing this easily all my life.

A few hours later, I lay in my bed, the man snoring beside me. I was nearly asleep when someone began knocking at the front door. The sound startled me. Maybe it's a mistake, I thought. I lay there, not moving, hoping that the person would realize it was the wrong apartment and go away. But as the knocks got louder, I became afraid. The man went on sleeping. Maybe he's married, I thought. Maybe his wife followed him here. My thoughts were wild, my heart was racing.

I didn't know what to do. I got up and walked into the hallway, naked. I thought about calling the police, but what would I say: someone's at my door and I'm afraid to answer? The streetlight was shining through the kitchen window, throwing my shadow across the wall. I looked at the black form of my body. I had never walked naked from one room in my apartment to another. I had never slept naked. The front door had a large pane of glass. The knocking became so hard I feared it would break. Maybe whoever was out there planned to break the window, unlock the door, and come in. And then the knocking suddenly stopped, leaving a silence that was just as frightening. I was still standing in the hallway when the phone rang. The man bolted up in bed. I ran back into the bedroom and picked up the phone.

"I knew you were home!" Michael said. "Why didn't you answer the door?"

"I was scared."

"Well, I'm calling from the corner. Let me in."

I put on my robe and went into the living room to wait. When I heard his footsteps in the hallway, I opened the door.

"Dad's dead," I said as soon as I saw his face.

He shook his head.

"Sean," he said.

By August, sitting beside the pool at the Holiday Inn, I had gained back all the weight I'd lost the previous winter. If I ventured into the bar, I would assume the invisibility of the obese. I lay back on the concrete, leaving my feet hanging over the side; gentle waves rose up around my ankles. The Ivory soap I'd stolen from the maid's cart spilled out of my pocket. I had forgotten about it. I sat up, unwrapped one of the tiny bars, and sent it floating away from me.

"Let's buy Ivory soap. It doesn't sink," I said, Mary and I tailing my mother in the IGA.

My mother didn't answer.

"It's impractical. It's soft soap. It doesn't last long," Mary said.

"Can I buy some myself?" I asked.

"If that's how you want to spend your money," my mother answered.

I bought the smallest size — "a beauty bar," Mary called it. I floated it every night in the bathtub and every morning in the sink. Soon only a sliver of soap remained.

Now I unwrapped all the bars and launched them from the side of the pool, moving my legs slowly through the water. The bars crested on the small current I created, bobbed up and down, and floated out to the middle. I watched the thin white rectangles drift together and apart, forming a kaleidoscopic pattern on the water's surface. "It's site-specific art," I said to myself, thinking of Ellis, whose specialty that was. She had loaded her car the week before and left for California, to the first of many teaching jobs that would take her from one art school to another until she got tenure somewhere and stayed. "Why don't you come with me?" she had asked. Watching her leave, I understood even better what I had begun to understand the night after Sean died, lying in bed beside Ellis. I understood how loss and love could situate themselves inside you, like one indiscriminate emotion.

And I understood too the nature of desperate acts. I could have gone with Ellis, easily, desperate for what we had to last a little longer, frightened that I would never again love or be loved. That was the desperate act — to live as if there were only one opportunity, as I had done the night Sean died, desperate to be desired by a man, thinking it was my only chance, saying yes, and wanting with that one word to leave behind all the years of never being noticed. Yes. It might have been what Sean was thinking when the coach finally called him off the bench the day he died, letting him play out the last few minutes of a basketball game his team was already winning. Yes!

Yes, a chance. And yes, he had embarrassed himself, making a mistake, being made fun of for it, then acting on desperation, leaving his whole life behind, certain that he would never be known — never be noticed — for anything else.

"No girl will ever like me," he had told me a few weeks before he died, after a dance that he'd deemed a "disaster." "Sure they will," I said, and when he asked me about my first school dance, I told him. The cutest, most popular boy had asked me to dance, and I was thrilled, thinking that he liked me. As we were dancing, the boy patted his back pocket each time any of his friends looked our way, signaling to them that he'd accepted a dare. "How much did they pay you?" I finally asked him. "Ten dollars," he said, proud that he'd gone for the biggest dare, the one that paid the most money. "I hate that guy," Sean said when I finished the story. "I hate that guy," he said when I told him about another boy, who asked me to leave a party early so everyone could play Spin the Bottle, saying that I was too ugly to kiss. "You're a complete cow," the boy told me, and when I finished the story, I saw that Sean was crying. I had been thinking of those boys, those humiliations, on the night the man picked me up at Dunkin' Donuts. From the moment he kissed me, I waited for the catch, waited to hear the laughter, to be let in on the dare, the joke. But there was none; there was only Michael, arriving after.

When I shared with Sean the stories of my adolescence, he said he hated those boys. But I didn't. I hated myself instead, just as Sean came to hate himself — in an instant, always. Yes, desperate acts. I understood them. I could have given him a detailed account. How to describe that night and the next one — the night Sean died and the night after. Coincidence one night, consummation the next, Ellis suggested. No. It was desperation one night, real desire the next. It was reverberation, the long end to my own adolescence. And then, rid of it, free of it, I was being who I was, leaving behind my fear and loving whomever I wished — a woman, Ellis. I was reaching out despite the risk, acting, finally, out of my own desire instead of waiting, wondering, hoping, at every dance, to be the one someone wanted.

There were other stories that, in time, I might have told him. In time, there might have been other stories he would have told me. But now there were no more stories for us to tell. There was no more talking to each other. "Let's have a conversation," he said to me once when I was giving him a bath. He was three, I think. I was twelve. "We're having a conversation," he yelled down to my mother when, from the bottom of the steps, she called up to see what we were doing. "We're having a conversation," I whispered, sitting by the pool in Chicago. I watched the soap floating out toward the deep end. In my lifetime, I have never been lonelier.

The next day, instead of taking the highway, I drove the long way home, going across northern Illinois and then following the Mississippi south to St. Louis. Sean's green jersey was on the seat beside me. Every few miles, I picked it up and pressed it to my face. It didn't smell like him. It didn't smell like anyone or anything. But I held on to it. I would hold on to it all my life, taking it with me wherever I moved, keeping it in a drawer with the rest of my clothes.

The last time I saw Sean wearing it was a few months before he died. I had stopped by my parents' house after supper. He was in the kitchen, washing the dishes. My mother was sitting at the table writing a letter. Sean turned from the sink when I walked in. He looked surprised to see me. "I was just thinking about you!" he said. "Just now, while I was washing this glass," and he held it up as evidence, as if his thoughts were reflected on its surface. "I was hoping you'd come over," he said, and he set the glass in the dishrack, put his arms around me, and hugged me more tightly than he ever had, sinking his head into my shoulder, his hands wet against my waist.

I caught my mother's eye. She had looked up from her letter and was smiling at us, and I could tell that she was touched, as

I was, by the tenderness of her teenage son. Sometimes, when I come across the green jersey, what I remember most strongly is Sean's embrace; other times, I see my mother's face.

I never found out why he was hoping I'd come over that night. When he hugged me, it felt as if he were relaxing into me, letting go of something, and I wondered for a long while what it was. After he finished the dishes, he snapped the towel against my leg before heading out to find his friends, and the evening went on like any other.

Driving home from Chicago, I stopped every few miles and cried in my car. By the time I reached St. Louis, it was after midnight, but despite the hour, I went back to my parents' house instead of going to my apartment. When I got up the next day, I sat on the steps in the front hall and looked out the window while my mother said her prayers in the living room.

"He didn't come yesterday," she said, as if she knew I was waiting for the blond-haired boy. Shortly after, she put away her prayer book and walked past me up the stairs. I sat there for a long while, hoping to hear something, the slightest movement coming from the second story, any indication that she had not gone back to bed, but the whole house was silent. I tiptoed upstairs. Kelly was still asleep, and my mother was in Sean's room, lying on his bed. In her hand was the letter he had left. I recognized it from across the room, his large handwriting filling the pages of lined paper. It startled me. It was months since I'd seen it. Had she been reading it all weekend? Written to a friend of his, it had been left on his dresser for someone — it would be my father — to find.

Dear _____,

I'm not sure if you'll ever see this, but I have to write it in case you do. I'm going to end it all. I'm going to kill myself. I'm going crazy, no insane. I've been contemplating this for a long, long time, off and on. I'm incompetent and incapable and I'm getting tired of trying futilely to accomplish something or be someone.

I want to be loved more than anything else and I have never

felt loved or wanted or needed. I tried to tell you a few times how I felt lonely, unimportant, and unwanted, but the problem is I can't force myself to say anything that matters. I try to bottle up all my feelings and not tell anyone what I really feel. There are times when I try to tell people my problems but I never really get too far. Tonight I called you on the phone. I wanted to tell you I was going to kill myself, but I couldn't.

I know what you are probably thinking. You are thinking that I am some kind of fool. You probably think I have it made compared to some people. I get straight A's and I was the best cross-country runner in the school. I make all kinds of money and have nice things and it seems like I'm good at a lot of things. Well, that doesn't matter to me anymore. What matters is that I don't like myself anymore. As a matter of fact, I hate myself. I'm not good at making friends or being one. I'm sure there are plenty of people who think I'm all right, but I don't think I am. I wonder how many people know of all my failures. I can't bear it. I take a lot of things too seriously I suppose and I guess I carry too much guilt. Anyway, I won't have to worry about it anymore.

You know what? Nobody talks to me at school after first hour. During third hour, the girls who sit behind me put paper wads down my back. They call my name and when I respond, they don't. They whisper about me and laugh and ask me if I'm going out with anyone. They have, on occasion, stuck signs on my back. I usually try to ignore them but today I turned around and one of them promptly said "Turn around, Sean, I don't want to see you." Then they both laughed.

When I eat lunch I sit with people who don't like me. Nobody talks to me. Nobody likes me in advisory I guess, and today I really messed up in science. I hate basketball and am no good at it.

The Christmas holidays really stunk, especially New Year's Eve. At twelve, I didn't even notice I was alone. The dance was a drag. I acted like such a fool. I hate myself for even dreaming that anyone would like me or go out with me.

I give up. I quit. This is the end. I wonder what hell will be like.

Love (if that's possible),
Sean

The letter dangled from my mother's hand as if she had just read it for the first time, as if she had been knocked down, flattened, by the weight of what he wrote. Had she been this way all weekend? I wondered. Had she been going back to bed? Was it the absence of Sean's clothes, his empty drawers and closet, that caused her to lie down, or was it the absence, yesterday and today, of the blond-haired boy? I stood in the doorway, watching, waiting, hoping she would ask what time it was. What time is it? she would ask me with her eyes closed when I was little, and I would tell her, and almost exactly fifteen minutes later she would get up. "Mom," I wanted to say, but I didn't. Instead, I went downstairs and set the souvenirs I had bought for Kelly on the kitchen table, along with my mother's list, and went home.

Each morning for a few weeks, my mother waited for the blond boy, but he never returned. The summer ended. Eight months had passed since Sean died, but I think it was only then that my mother began to feel his loss most fully. The blond boy was gone. Sean's clothes had been given away. A new school year began without him.

Ten years later, my father phoned me in New York to say that the blond boy had come back. It was a Saturday in early September. Out washing his car, my father noticed the boy heading slowly up the street. Since it was already afternoon, there were no newspapers to pick up, but he stopped at each house and scanned every lawn. When he got to our house, the boy walked up the driveway, said hello, and asked my father his name.

"You're Tom," the boy said after my father told him, and then he asked when my father's birthday was. "You're Tom," he said again. "You're born October fourteenth," and he walked away, repeating this newfound information over and over until, when he was halfway up the hill, my father could no longer hear him.

"'You're Tom. You're born October fourteenth,'" my father kept saying. "What do you think of that?"

The boy looked the same as he had the last time he had come to our house. According to my father, he hadn't grown or aged. A few weeks later, my mother was sitting in the living room saying her morning prayers — again, it was a Saturday — when she experienced what she described as a terrifying sensation.

"It felt like someone had come up to the house and put their hand on the door, ready to come in, but when I looked out, the porch was empty," she said. "I sat back down but I still felt someone's presence; I can't explain it. And when it was gone, I felt this sadness come over me, this huge loss like something had been taken away from me."

Later that afternoon she noticed that one of the statues my father had bought her the summer after Sean died — the one that looked like Sean, she said — was gone. "Someone came right up to the house after all these years and just took it," she told me.

Though there was no reason to suspect him, I couldn't help thinking of the blond-haired boy, and I kept picturing him as he looked the day I drove Sean's clothes to Chicago, playing dead on the driveway, his eyes fixed on the front porch. No one ever learned who he was that summer. No one knew where he had come from or where he had gone, but I had convinced myself that my encounter with him had caused his disappearance. If I hadn't "hit" him, I thought, he would never have left. In my mind, I had collided with the mystery and made him vanish.

Now, a decade later, he had appeared again. Would he return every ten years? Would he come back on my father's birthday? "You're Tom. You're born October fourteenth." Was it information he was after, or — in place of picking up papers and putting them back down — was he looking for pleasure in a few words he could remember and repeat? I kept thinking about him, wondering who he was, until my mother, still talking about the statues, reclaimed my attention. "They only took Sean," she said. "They didn't take Kelly."

"Or the Virgin Mary," I added. "Or the mushroom, or the

dog that looks like Dad, or the frog, or the flower cart, or the squirrel."

"It's not a squirrel, it's a chipmunk," my mother corrected.

"Or the chipmunk," I said, and I continued to list the statues that, over the years, had proliferated across my parents' front lawn.

"No, none of those," my mother admitted, and she laughed.

"Of course not," I said. It was only Sean who was missing.

New Year's Day, 1990

It is New Year's Day, just turning dark, and my father is walking from room to room waiting for the rest of the family to arrive while my mother sits at the kitchen table painting her nails.

"Reebop!" my father says as he comes into the kitchen. "What do they call that color, Lois? Red?"

"You're not going to start that tonight, I hope," my mother says.

"What would that be?"

"That reebop business."

"I gotta give those kids what they wanna hear. Makes 'em just go nuts," he tells me.

"Yeah, I've heard," I say. I am sitting across from my mother, reading the paper.

"Oh that got back to you, did it? Yeah, they think their old grandpa's crazy, plumb out of his mind. ReeBOP!"

"That's what they think, all right."

"What'd they tell you?" my father asks. He is annoyingly energetic.

"They asked me, 'Why does Grandpa always say reebop?'"

"And what'd you tell 'em?" he says, drumming his hands on the table.

"Tom, don't shake the table," my mother says without looking up from her nails.

"I told them it's because you forgot the words to 'Can I sleep in your barn tonight, mister,'" I say, referring to the song my father used to sing, nonstop it seemed, through his thirties and forties. ("Does your father still sing that song?" my friends ask me.)

"You all have never appreciated my talents," my father says, and he starts to serenade us.

My mother laughs. "See what you started?"

Every holiday begins the same way at our house. Everything is ready and waiting. The dining room table is set. Whatever main dish my mother has made is in the oven, and on a section of the kitchen counter my father has set up an ice bucket, glasses, and the bottles of wine, soda, and liquor he will use to make drinks. On the coffee table in the family room are crackers and cheese, chips and dip, nuts, slices of sausage, and three or four kinds of candy.

"Watch the kids with that candy," my mother will say at some point. "I don't want them taking chocolate into the living room."

This is how the last minutes pass before everyone arrives. Mary will show up first, with a salad. Then Kelly will come, with the latest rage in Jell-O dishes. One year a key component is crushed potato chips, the next year Frosted Flakes, another Ritz crackers. Each year she reveals this believe-it-or-not ingredient with the same enthusiasm as the year before, and each year my mother will seek clarification. "Any kind of crackers, or do they have to be Ritz? Any kind of frosted cereal, or does it have to be corn flakes?" Michael will arrive last, his wife bearing a vegetable.

These are the things we can count on. And always, as a prelude, my father will walk the rooms, repeating whatever word or phrase has popped into his mind and stayed there, a kind of comic mantra that has become a central part of his personality as a grandparent.

"I guess you think I'm crazy too, huh?" he says. He is next to me now, doing some kind of dance step he's seen on TV and tapping my arm with his fingers.

"It occurs to me from time to time."

"You people don't understand what I'm up to. You can't follow me. I'm too far ahead," he says.

My mother smiles. "Well, what is it you're up to?" She is shaking her hands now, each of her nails shining a wet red.

"Shake it out, sister. Shake it on out," my father says. "Praise the Lord and Happy New Year."

My mother laughs. "This is how he is now. Count yourself lucky you don't live here anymore."

"ReeBOP!" my father says, and he opens the refrigerator and stares at the shelves.

"What are you looking for?" my mother asks.

"I'm just a man lookin' in his own refrigerator," he answers.

"Well, do something useful and put this polish away for me," she says. She is alternately blowing on her nails and shaking out her fingers.

"Yes, ma'am. Got that," my father says, and he snaps the polish off the table and sets it beside the other bottles of nail polish on the top shelf of the refrigerator door. "We gotta cool down that ruby red. Yes, ma'am. We gotta reebop!"

My mother rolls her eyes at me. "Here's to another high holiday," she says, and she tells my father to fix her a drink.

Soon we hear Mary's family at the door, and as we move from the kitchen to the front hall, my mother says, "Tom, I mean it now. Turn it down a notch or two before the kids come in."

Mary goes right to the refrigerator and clears a space for the salad. "I hope this is enough," she says. Our family has been the same size for a while, but still she says this every year. "How are you? Happy New Year," she says, and she kisses my cheek. She is all cold winter coat and perfume and reminds me of my mother.

"Who's here? Who's here?" my mother says, clapping her hands and clasping them at her chest. "Allison! Come give me a kiss."

"Happy New Year, Mom," Dan says, barreling through to the kitchen with a six-pack of beer. He is the only one of her children by marriage to call her Mom. It is the same for my father. "Happy New Year, Dad," Dan says, as he heads for the refrigerator.

"Dan, I got beer here. Why'd you bring that?" my father says.

"I don't know," he says, and he sits down, beer in hand, next to the plate of sausage in the family room.

"Jesse! You look so handsome!" my mother says. She is still in the hallway, kissing Mary's kids. My seven-year-old nephew has a new haircut. It is short on the sides, long on the top, and parted in the middle. "My Uncle Mark owns his own beauty shop," he tells us, referring to Dan's brother, who has just been up from Florida for a visit.

"C'mere, kid," my father says, "and give your old grandpa a hug." Jesse puts his arms around my father's waist and presses his head against my father's stomach, closing his eyes and smiling, as if he intends to settle in and stay there. Of Mary's kids, he is often the most affectionate. After a moment, my father takes Jesse's face in his hands and says, "How much your Uncle Mark charge a man for a haircut like that?"

"I don't know."

"Well, tell'm he owes your old grandpa a free one for lettin'm hang around my house when he was in high school."

"Unh-unh, Grandpa. He lives in Florida."

"He does *now*," my father says. "Hightailed it down there and left the door wide open here for your dad. That's the way that went down. You got me, kid? Reebop!"

"Grandpa!" Jesse groans.

Sarah, now eight, has the same kind of beauty Sean had, with an abundance of blond hair and flawless skin.

"Happy New Year, Sarah! I'm so happy to see you!" my mother says, wrapping her arms around her oldest grandchild.

This all happens in the hallway. It is a crowd of kisses and winter coats. On either side of the hall, the living room and dining room are lit dimly, and in the family room the fireplace emits the faint glow of the sunlamp Mary used to tan her face when she was a teenager. It is clamped to the damper, sending its simulated rays up into the chimney.

"They'll never leave now," Dan says. He is talking about the birds that have been nesting in the chimney since the previous summer.

"Well, I don't want them to leave," my mother says. "If they leave now, they might not make it. I thought they'd fly away for winter, but they never did. Now I don't know if they're stranded or if they know what they're doing and just decided to stay." She offers this explanation each time we're together, but the sunlamp is a new addition to the story. "Anyway, they're singing again, so they must be happy," she says, referring to the source of artificial warmth she directed my father to install after a day when the birds were uncharacteristically quiet.

"A sunlamp? I never heard of that," Jesse says, trying to peer into the chimney to see it.

Soon after, Kelly and Duane arrive, then Michael and Sauni, and the sunlamp is explained a second time and a third.

My father prides himself on knowing what each of the kids likes to drink. Orange soda for Allison, Dr. Pepper for Jesse, root beer for Sarah. "Reebop!" he says as he hands them each a glass.

Sarah is sitting next to Sauni on the sofa. "Do you know why Grandpa keeps saying that?" she asks my sister-in-law.

"Why?" Sauni says.

Sarah shrugs her shoulders. "Maybe because he's an old man."

Michael drinks Diet Cherry Coke. Sauni likes white wine. Mary takes a bourbon and seltzer. Duane joins Dan with a beer, and my mother and Kelly drink frozen strawberry daiquiris that have been in the freezer since Christmas. My father has an almost constant cup of instant decaf going, and I'm on my third or fourth Diet Pepsi of the day. We will all stay in the family room until it's time for dinner. "It won't be long," my mother tells us, taking her daiquiri into the kitchen.

Kelly has a white rabbit with a patch of black fur over its left eye. Though Kelly and Duane live elsewhere, the rabbit remains in residence at my parents' house, in a huge hutch in the back yard. A condominium, my father calls it.

"Can we bring Domino in?" Kelly asks.

"Yeah. Can we? Can we?" the kids ask, echoing each other.

Kelly brings the rabbit in, and it sits motionless in the center of the room, the kids' legs spread out to form a triangle around it.

"He has to stay inside our fence," Allison says when my mother comes in from the kitchen.

"Ohhh! No, he doesn't," my mother says, and she picks up the rabbit and sits down in the rocker. "There's my baby," she says, nuzzling it against her neck. "Come here," she says to Sarah, Jesse, and Allison. "You have to hold him on your lap and let him know you love him."

The kids sit in a line on the floor in front of my mother. She is a short, round woman with snow-white hair, and as she rocks forward to put the rabbit first in Allison's lap, then Sarah's, then Jesse's, Michael says she looks more closely related to the rabbit than she does to any of us.

"Go ahead and tease," she says. "Tell him, Allison, he's an old tease."

"I can't imagine what a mess it must be up there," Michael says when the birds chime in for the first time that night and fill the fireplace with music.

"Oh, they're not hurting anything," my mother says. "They're just a few little birds." And she gives up the rocker to Kelly and goes back to the kitchen to check on the meat. Mary follows to help, and Michael looks at my father and laughs. "It sounds like we're sitting out in the back yard," he says.

My father smiles. Implied in Michael's statement is the question of why my father hasn't done anything to get rid of the birds. When it comes to the house, it is unlike him to let something like this happen. "Well, y'know, we never use it anyway, and they make your mother happy. What can I getcha, Sauni?" he says, reaching for my sister-in-law's glass.

Just before dinner, my mother takes Mary and Sauni upstairs to see her latest arrangement of silk flowers, a kind of linear wreath, horizontal rather than round, that hangs above my father's bed.

"Up to see the flowers, are they?" my father says when he comes back up from the basement, where he has gone with Jesse to get a box for the rabbit to stay in while we eat dinner. "I'm under a lotta pressure here," he tells us.

"Why's that?" Michael asks.

"Your mother's got me sleepin' under a funeral spray," he answers.

"It's a swag, Dad," Kelly corrects. She has taken the same class in flower-arranging as my mother.

"Oh, a swag, a swag," he says. "Well, all I can say is it's the perfect size for a coffin. Gets me a little nervous goin' to bed."

When it's time for dinner, Kelly, Duane, and I eat in the kitchen with the kids. "You don't mind, do you?" my mother says.

"Is everyone all right in there?" Mary calls every few minutes from the dining room. "Are they eating?" she asks.

"Kelly isn't," Allison answers. "She says she only likes Jell-O, Mom."

"That's okay, Allison. You eat," Mary calls back.

After dinner, my mother wants us to watch a video. It will be her first viewing of a gift she received for Christmas from her brother, who has begun to transfer his home movies to tape. She has been waiting all week to watch it.

"Will I be in it?" Jesse asks.

"No," Mary explains. "Uncle Walter quit making movies way before you were born."

"I guess I won't be in it either," Kelly contends.

"I doubt it," Michael says, "I think he quit making them even before Sean was born."

"Figures," Kelly says, and she crosses her arms and cocks

her head, assuming the pose she always takes when she's pouting.

All I remembered about the movies my uncle made were rooms flooded with lights when we were little. I remembered the whirring sound of the lights as they got going and the sudden heat they emitted. I expected to see rooms full of people sitting around squinting.

The video alternated between scenes of summers spent at my grandparents' house in the country and scenes of Christmas, charting the growth of our family at six-month intervals. We were always the same group — my grandparents, aunts, uncles, and cousins on my mother's side — and almost every year the group grew larger, with a new baby being passed from person to person. By the end, ten children would be born to my mother, her sister, and her brother.

The first time my grandmother appeared, she was riding her old white horse and waving. Seeing her, I understood more fully the meaning of the term "moving pictures." Before, we had had only photographs to remember her by — static, unanimated images. To see her moving was to see her alive again, resurrected in a way, indifferent to death as she rode to the pasture, got down off her horse, waved again, and walked out of view. It was the same for my grandfather, a man of whom my only memory was through photographs. He was moving, walking from the bench where he sat shucking corn with my uncles to the big stone barbecue pit where he had four slabs of ribs going on the grill. He was tall and thin, wearing only his swim trunks and sandals. So that's what he looked like alive, I thought as I watched him spear the meat, somersault it onto the grill again, and walk back to join my uncles.

In the movie, we were all moving. It wasn't just the way the dead were moving among us. It was the way we moved as children. It was the way our parents moved before we were born. Early on, in a scene in the country, my mother and father are swimming in the river up the road from my grandparents' house. "The Meramec," my mother says when Sarah asks. My father is floating in an inner tube, and my mother is next to him, treading water. Hanging over the inner tube, my father's legs look more muscular than I have ever seen them. Suddenly, my mother lurches forward, lays her body over my father's chest, and kisses him. His feet fly up out of the water, and we see that he's wearing brown leather dress shoes and black socks.

I don't remember my uncle filming us in the country, probably because the scenes are all outdoors and required none of the blinding lights that remain my only real memory of my uncle making movies. The scenes from the country are among my favorites. There is a shallow plastic swimming pool near the pasture. At best, it is a foot and a half high and five feet wide. My father and uncle take up most of it, with Michael, Mary, my cousins, and me falling in all around them. In the yard, a wooden porch swing sits between two shade trees. My mother is sitting on one side of the swing, holding me in her lap. My aunt is on the other, holding my cousin. Mary sits in the middle. My cousin, Mary, and I are wearing only our underpants. I am three, my cousin four, and Mary five. My mother and my aunt are talking as they idly keep the swing in motion. We cannot hear what they are saying. The movies my uncle shot are all silent.

Watching myself learning to walk, it's as if I am seeing someone I never knew. As I stumble through, the men sit plucking chickens on the bench near the barbecue pit, a pile of feathers forming at their feet, while in the background my grandmother, a flapper in the 1920s, teaches my older cousins to do the Charleston. They are all moving, my grandmother's movements more familiar to me than my own.

In the country, there's a bottle opener mounted on the side of the house, with a bucket full of bottle caps below it. The men collected them for a game, and in one scene we watch as my mother's favorite uncle, Ben, pitches bottle caps to my father, who stands ready to bat with a broomstick. In the summer that follows, Ben is nowhere to be found. He is the first of the missed to be missing.

One year the house itself will be sold, and the rest of the video shows us only at Christmas.

Every Christmas Eve, until we were too old to fool, my father, my grandfather, and my uncles drove all the kids around in the car after dinner while my mother, my grandmother, and my aunts spread the presents under the tree. We were supposed to believe, and for years we did believe, that Santa Claus had come while we were away. This ritual was represented on film by a shot each year of Michael, Mary, my cousins, and me coming one by one down the steps to the basement of our house, or our uncle's, or our aunt's, a look of anticipation and then surprise on our faces when we reached the bottom and saw the presents. Seeing these shots, I am moved by the tenderness of the men as they hold our hands down the steps, help us unwrap our presents, and show us how to work our toys, our uncles treating us as if we were their own children.

And it is the Christmas scenes, unlike those in the country, that show the growing prosperity of our parents. In the early Christmases, we are seen celebrating in unfinished basements. The grown-ups sit on folding chairs in a circle in front of the furnace. A small Christmas tree has been set up on a table, and paper decorations hang on the concrete walls. As kids, when we came down the steps, we were likely to be wearing clothes that one of our cousins had worn the Christmas before. One year my oldest cousin is wearing a red taffeta dress. The next year it is Mary's. Then it is another cousin's. And another's. Then mine. By the third sighting, we begin to joke about it. Who will come down the steps this Christmas wearing the red dress?

"I guess it never occurred to anyone to save it for me," Kelly says.

"Well, since you're not in this, I guess we'll never know," Michael teases.

"Michael," Sauni says, "you're not helping."

"I have a red dress," Allison offers, and even Kelly softens.

As the years go by, the basements get finished. The walls are paneled. We sit on sofas and upholstered chairs. In our basement, there is a brick bar; in my aunt and uncle's, a wooden one with a mural of a mountain behind it. The Christmas trees are no longer tabletop size. They stand on the floor and reach the ceiling. The men are heavier and act more like spectators than participants, though they are still solicitous toward us, and free with their affection. Our mothers look more experienced, like grown women. Yes, they look more like women, and the men look more like men. I glance at Kelly, who is trying not to act too interested in what we're watching, and I realize that she is about the age my mother was when the video started. I look at Mary, about the age my mother is at the point we have reached, and I see that it is this span of so many summers and Christmases that separates Kelly from Michael, Mary, and me, the basement unfinished to the basement finished, the starting out to the settling in, and I suddenly wish, for Kelly's sake, that she were in it, that my uncle hadn't stopped making his movies before she and Sean were born, that she would walk down the steps wearing the red dress.

But it is before everyone comes of age, before the traditions become tradition and the starting-out years yield to something like stability, that my mother is moved to tears. One Christmas begins with a closeup of my grandfather. His hair is the same snow-white as my mother's is now, and he is wearing a pale blue shirt. In his arms, he holds a baby in a pale blue dress. It is Mary. She is one month old, the first baby to be born into the movie. I look at my mother. Tears are streaming down her face. "They're both wearing the same shade of blue" is all she says.

That Christmas, the camera follows Mary as she is passed from person to person. She looks at ease with everyone, as impartial as the black patent-leather purse that was passed from person to person before her, a present my grandmother received. Everyone held it up high by the strap, its clean, simple shape connecting one person to the next, like the paper-doll chains my mother taught us how to make when we were little. There was a kind of unconscious consideration with which this act was carried out, everyone handling the purse by the strap so as not to smudge the patent leather, until the purse came full circle and my grandmother placed it back in its box.

It looked to me like love, like blessedness and beauty, the black patent-leather purse, and, just after, Mary making her way around the same circle in her pale blue dress. Mary, as I have never seen her, alive before I was born.

"Everyone wants to hold you, Mom," Sarah says.

I want to hold her too. "Replay this part," I tell Jesse, who is manning the remote, and we watch Mary being sent high-speed around the circle in reverse until we catch the first sight of her again, in her blue dress, asleep against my grandfather. She is in my aunt's arms when I realize what I am feeling. It's what I felt the first time I saw Sean. He looked familiar, as if I had known him a long time already. I remember the moment clearly, when I bent down to kiss him as my mother told him my name. I was overtaken by a strange sensation, a kind of visceral realization that time had tenses, that I had a past, a present, and a future. I remember the confusion it caused me. Was this how it felt for everyone, or was I reacting to my new role in the family? He was the baby now, and I was not. I was eight, going on nine. I had no answers. But I remember being unable to stop thinking about time, as if Sean had somehow transformed it. It must have shown on my face.

"What's the matter?" I remember my mother asking.

"I don't know," I told her. "I feel so different."

"Good different or bad different?" We were in the hospital lobby, and she was sitting in a wheelchair, holding Sean. I remember her hand moving from Sean's stomach and slipping into mine.

"Good different or bad different?" she asked again.

"I don't know," I answered. "Serious different."

"Serious different?" Mary said. Michael had gone with my father to get the car.

"Like time is different," I said, and shortly after, a few weeks maybe, I remember walking up and down the street with Mary. It was raining. We had my father's big black umbrella and we were pretending we were a married couple walking in the rain. "Fred and Ethel Mertz," Mary said, naming the neighbors on our favorite TV show, *I Love Lucy,* and she began singing "We were walking in the rain one day, in the merry, merry month of May" in a loud, deep voice that seemed to challenge the very weather around us. It was late October or early November, a Saturday, and no one else was out.

"You're awfully quiet today, Ethel," Mary said. She had taken the part of Fred because she hated to walk under an umbrella that someone else was holding. "When a man and a woman share an umbrella, it's the man's job to hold it," she said. I knew that already, but I didn't bother saying so. I just put my arm through hers and played the part of Ethel. We walked to the end of our street, crossed, and walked back along the other side. Mary was singing and talking like Fred the whole time, but my thoughts were elsewhere.

"Ethel, sing or say something," Mary said, nudging my arm.

"Do you ever think about what it's like to not be alive?" I asked. Mary had begun whistling, but when she heard my question, she stopped.

"You mean, to be dead?"

"Yeah. Well, to be dead or not to be born," I said. It was Sean's birth that had started me thinking about this, the fact that he existed now and hadn't before. Sean's birth, and something we were studying in school about souls. I couldn't comprehend nonexistence, and I kept coming back to it whenever we were allowed to ask questions in religion class, until the teacher sent a note home and my mother asked me to stop.

"I never thought about it," Mary said. "Why do you need to know?"

"I don't know. I keep trying to figure it out, but nothing happens. My head just gets empty. I can't picture what it's like not to be alive."

"Well, can't you think about it later?" Mary said. "It's raining right now."

Rain was Mary's favorite kind of weather. She loved to be

out in it. She loved to play in the puddles it left on the street. The section in front of our house was prime property; any puddle that formed there was usually so wide and deep, it took several days for it to disappear. Mary washed her hair in it once. She was only six or seven, but she had already developed habits aimed at improving her appearance: brushing her hair a hundred strokes before she went to bed; keeping her skin well lotioned in winter; patting her face dry rather than rubbing it. Beauty secrets, she called them. So when she heard that rainwater would bring out the natural shine in her hair, she was more than willing to try it. She squatted down and dipped her head in the puddle, but she had barely started when my mother came out, pulled her into the house, and soaped up her hair in the kitchen sink. Years later, when we were teenagers, she would finally triumph in her efforts. She collected rainwater in buckets in the back yard and poured it into empty shampoo bottles that she saved under the bathroom sink. Sean set out buckets in the back yard too. While Mary harvested rainwater for her hair, Sean collected it to fill his aquariums, claiming there was something beneficial in it for his fish. Mary got her information from magazines and from listening to other girls talk about beauty. Sean got his from a book he bought called *Raising Fish for Fun and Profit*.

But beyond their utilitarian appreciation of it, there was something about the rain that made Sean and Mary happy, and Mary's exuberance often manifested itself in the games we invented, like this one, in which we played Fred and Ethel Mertz, strolling up and down St. Patrick Lane, singing. (It was not St. Patrick Lane, Mary would say. It was Fifth Avenue or Park Avenue or Broadway or one of the other New York boulevards famous enough for us to have heard of in the semirural suburbs of St. Louis.)

The truth is, it was in the rain, alone with Mary, that I felt the closest to myself, to the notion I had of who I was and who I would come to be, even if I was pretending not to be myself, playing the part of Ethel Mertz to Mary's Fred, or Debbie Reynolds to her Gene Kelly. And it wasn't just because Mary took the man's part whenever it rained. (In the rain games, there was no sex. There was just the man's superior skill at leading the singing and dancing and holding the umbrella.) None of that really mattered to me. It was being with Mary when all her defenses were down, as if the rain washed away the outer edge of her personality. It was something, as an adult, I would come to associate with every cold gray day — the comfort of Mary, the closeness of singing with my sister. It was the safety of being next to her, my arm in her arm, rain all around us, the world as we knew it lined up house by house on either side of the street, the people inside the houses seeing us through their windows, recognizing us together, knowing who we were, the two of us out in the world, in the cold close comfort of rain, out in the world we created with each other, Mary for me, me for Mary.

"Why don't you wait and think about all that another time," Mary suggested. "I'll help you think about it later if you want. When we're back inside. Or tonight, when we're in bed. Okay? C'mon. It's raining."

"Okay," I said.

"Good. Then play your part." And she started singing the song again, her own version of something she had heard Fred Mertz sing in an episode of *I Love Lucy*. Of all the men on television, she loved Fred Mertz the most.

"We were walking in the rain one day, in the merry, merry month of May," she sang. Mary had strawberry-blond hair. My hair was blond too, but not the same shade as Mary's. It matched the color of the water in the kitchen sink. Dishwater blond, my mother called it. My hair was stubborn and straight, and Mary's was wavy. She had hair that could be styled, my mother said. I had hair that had to be trained. It was no wonder my mother pulled Mary from the puddle. Even pretending to be Fred Mertz, Mary was beautiful, and though I never knew Fred and Ethel to be affectionate, I rested my head on Mary's shoulder and walked with her that way in the rain. "Sing something else, Fred," I said, touching her sleeve.

"Oh, criminy, Ethel," Mary said, "what d'ya wanna hear?" But before I could answer, she began belting out "I'm looking over a four-leaf clover," and we continued walking up one side of the street and down the other until it was time to go in for dinner, Mary singing every song she knew, her false tenor registering against the rain.

"Again?" Jesse asks. "Do you want to see my mom being born again?" He is standing ready with the remote.

I laugh, amused at his choice of words. A few years later, I would watch the videos of each of Kelly's kids being born. Kelly in the delivery room, her legs spread open, the appearance of a head, a body, the cutting of the cord. Real as it was, it didn't seem real to me, and the impression it made on me — as much as I love her, as much as I love Stephanie and Nick, my youngest niece and nephew — didn't even come close to the feelings I had seeing Mary on film as an infant.

She was born wearing a pale blue dress. That's how it would end up in my mind, the enduring image I would have of her as a newborn, as real to me as the fact that Sean entered the world with a bruise above his eye. And what was the connection between seeing Sean as a baby and watching Mary in the video? As babies, their faces were already familiar to me, as if I had seen them both before. It was as if we had each left some presence of ourselves alive in my mother's womb, some memory, some essence that allowed us to recognize ourselves in each other, a kind of prenatal knowledge linking the child already living with the one about to be born. There was something about the connection, about seeing some aspect of myself in the people born before and after me — in Mary, in Sean — that made me feel immortal, and I wondered, watching the movie, whether Sean ever felt that I was his protection, as I always felt that Mary was mine, convinced that as long as she was alive, I would be also, that I would never be sick or destitute or completely alone as long as she was living, as if her existence guaranteed mine. Was it something basic, something biological, or did it come from sharing a childhood, from sleeping together, those early bonds defying whatever distance had come between us as adults? And would Sean have felt the same for me had he gone on living? In this construction, this order of births, the person I was meant to care for, whose existence I was meant to guarantee, had been ripped away from me, causing time to collapse, the present pushed back into the past, the future foreshortened.

"Do you want more soda?" I hear Michael saying. He is standing next to me, his hand on my glass.

"Do you want me to wait for you, Uncle Michael?" Jesse asks. He has stopped the video on a shot of my grandmother dancing in front of the Christmas tree.

"No, go ahead," Michael says, and Allison follows him into the kitchen with her cup, saying she's thirsty for something, she isn't sure what.

"Milk, Michael," Mary yells after them, and my mother starts shushing everyone.

"It's a silent movie, Mom," Kelly says.

"Shhh anyway."

"Shush at the chimney, why don't you — the birds are making most of the noise," Kelly says, and she rolls her eyes and pumps the rocker harder. Duane, Kelly's fiancé, is sitting on the sofa next to my father, and in a low voice my father starts to tell him how Kelly wore a hole through the family room floor by rocking so fiercely when she was a teenager. She sat in the rocker for several hours every day, her headphones plugged into the stereo in the corner.

Kelly shushes them loudly. "Don't listen to him, Duane," she says.

A Christmas comes when my grandfather is missing. It takes us a while to notice that he's not there. In fact, we don't notice it until my mother says, "This must be the first year that my father's gone," and then his absence becomes obvious.

"Where'd he go?" Allison asks.

"He died, Allison," my mother answers.

"Oh. When did he do that?"

"Allison, come sit with me," Mary says.

"Which man was he, Mom?" Allison asks as she climbs onto Mary's lap.

"The one with the white hair," Mary tells her, and a few Christmases later, we hear Allison whisper, "Mom, if Grandma wants to see that man again, she should just tell Jesse to rewind," and shortly after, she falls asleep.

Sarah too has fallen asleep. She is wearing white tights and the green velvet dress she wore for Christmas, and watching her sleep so soundly in the middle of the room reminds me of her first Christmas, the Christmas that would be Sean's last. They had fallen asleep together on the living room floor, where he had taken her to look at the Christmas tree lights after dinner. "Sean and Sarah are asleep," my father told Mary when he walked through the kitchen on his way to the garage. Mary went to the hallway with a towel and the plate she was drying and peeked in at them. My father returned with a load of firewood in his arms. A shot of cold came through the kitchen door. "Did you see 'em?" he said.

Now, on a shelf above where Sarah lies sleeping, a photograph of Sean looks out at us. Shortly after he died, my mother took out all of the cardboard backing in a picture frame and filled the space with an entire set of his school pictures, from kindergarten through ninth grade. Every so often, she changed the order of the photographs so that a new picture of Sean was displayed. One month we might see him as a second-grader, wearing his First Communion suit. The next time we looked, he was fifteen, with his hair parted in the middle and his face showing its first signs of manhood. Or he was ten, looking somewhat timid, with an almost unnoticeable gleam in his left eye, where his tear duct, severed by a falling tree limb the previous summer, had been temporarily repaired with a tiny plastic tube that kept his eye from constantly "crying." The tube emptied his tears into his nose instead, so that, with his nose constantly running, it seemed as if he were always catching a cold. Then he was five, smiling for his first school picture, happily unaware that his teeth were abnormally tiny and congenitally discolored, little glints of gray along his gumline. When she was two and just talking, Sarah would point at each new picture and ask, "Who's that boy?" "Who's that boy?" "Who's *that* boy?" until she had grown used to the sight of her uncle from the ages of five to fifteen.

With Sarah and Allison both asleep, Jesse is the only grandchild awake. "There's Al," he says, pointing to Mary when, in the video, she's the same age as Allison. They look almost identical.

"That's not Allison, that's your mom," Michael says.

"No. It's Al," he says, convinced that it's his sister on the screen.

"It's me, Jess," Mary says, and he looks closer.

"That's right. Al doesn't have that toy," he says of the zebra Mary is riding. Its legs look like barbershop poles when Mary wheels toward the camera, the stripes spiraling down, then up, then down again.

As we watch, there are moments when we can see Mary's kids in each of us. A gesture of Michael's reminds us of Jesse. Something of Sarah shows up, briefly, in me. We see Mary in my mother, Michael in my father. One year, I look a lot like my aunt. Christmas by Christmas we see ourselves: Michael, Mary, and I heading toward adolescence, my parents toward middle age.

Michael turns into a teenager. "Boy, you just did it all, didn't you?" Sauni says as we watch him, dressed in black pants and a white turtleneck, doing the Monkey with my cousin David. It's the third time on the tape that he's shown dancing.

"Yep," Michael says, and as he starts to say more, the screen suddenly goes blank, and the next thing we see is Sean.

He is in my mother's arms.

My father lets out a loud cry that sounds nonhuman, the piercing call of a big wild bird, a crow cawing, and Dan looks first at the fireplace and then at my father as my father springs up from the couch and leaves the room in what seems like a single motion.

Kelly stops rocking and sits up straight. "Oh, my God," she says, and tears start streaming down her face.

Sean, bundled up in blankets, is being passed from person to person.

I look at Michael's face. Tears are rolling out from under his glasses. Mary has most of her face buried in Allison's head, Allison's limp body lying against her. I feel I can barely breathe.

"I thought you said Uncle Walter stopped filming before Sean and I were born," Kelly cries, and she leaves the rocker and climbs on the loveseat to sit next to my mother.

"I guess he didn't," my mother says, letting Kelly snuggle in close. "I guess there's some of Sean," and though at some point my mother too begins to cry, she seems at first to accept Sean's appearance more easily than the rest of us.

Somewhat recovered, my father returns and watches from the doorway. "Back it up a bit, will you, Jess?" he says. Sean is asleep through most of his first Christmas. The next year, he is walking. His feet are moving. His head is turning. His hands are picking things up and putting them back down. His mouth is opening and closing. He is laughing and smiling and talking to Michael. He walks over to my father, who puts his cigarette in an ashtray, slides it across the table, and picks Sean up. Sean sits on his lap a while, gets down, and walks over to Mary, who carries him upstairs. Soon after, he reappears in his pajamas.

Near the end of the video, in the final scene of Christmas, we see a baby sleeping on my mother's lap. "See, there you are," my mother says to Kelly, who, instead of smiling, rearranges her legs a little on the loveseat.

That year, the year of Kelly's first Christmas, my grandmother was given an ottoman. In the video, my uncle sets a big box before her, and Sean helps her tear off the wrapping paper.

"I remember that," Mary says.

My grandmother looks delighted. She is smiling at Sean, telling him, it seems, how much she likes the gift. Sean nods in response, claps his hands, and climbs up on the ottoman with my cousin Robert. Sean is three, Robert eight or nine. The ottoman is on rollers, and the tape ends with the two of them sitting on it, spinning in circles, each alternately whirling into view and whirling out again.

When the video is over, we sit facing the blank screen until Michael rewinds the tape to the scene of Sean on the ottoman, trying over and over to stop the motion each time Sean's face spins past. I understand what he's trying to do. In the years that follow, whenever I watch the video, I will try to do it many times myself, to freeze Sean's face in its full happiness. Always, I will fail. After repeated attempts, Michael gives up and rewinds the tape to the first sight of Sean, and we watch him again, one or another of us saying, "Rewind it," each time the tape ends, until, close to midnight, my mother suggests that we save the rest for the next time we're together.

Kelly looks confused. "There is no more, Mom," she tells her.

My mother nods. She has been quietly crying, and she gets up and goes to the kitchen. We hear her putting away a few dishes, and then she goes upstairs without saying goodnight.

Sarah, Jesse, and Allison are asleep, and Michael helps Dan and Mary carry them to the car. After everyone has gone home, my father goes upstairs too, telling me to turn off the lights. Home for the holidays, I am spending the night.

I lie on the floor in front of the fireplace. The birds have been quiet for hours. The sunlamp is still lit. When I wake up, I see my father sitting on the sofa in his robe. I wonder what time it is, but the clock on the wall isn't running. It used to chime on the hour, but months ago — the birds in early residence — my mother removed the batteries.

"What time is it?" I ask.

"I couldn't sleep" is all he says. The TV is on, no sound, and he's holding the remote. From where I'm lying, I can't see the screen.

Not knowing whether it's late night now or early morning, I listen to the silence, the faint warmth of the sunlamp filtering down through the fireplace. I wonder what time the birds will wake up. I see that my father has a cup of coffee on the table near him, and as I lie there, I realize that he's watching one part of the video again and again, the same segments of time elapsing as the tape runs and rewinds and runs and re-winds. As I listen, I think of Sean on the ottoman, his face wide with laughter, and that's what I expect to see when I finally get up and lie down next to my father. Instead, I see my father floating on the inner tube. My mother kisses him, and his feet fly out of the river, revealing his brown leather shoes and black socks. There were snakes in that river, and all kinds of bottom feeders, he had told us earlier, in his defense.

I lie next to him for a long while, watching in silence as he keeps playing that scene over and over, until, morning rising, he hands me the remote. "I was a young and powerful man" is what he finally tells me.

"Wasn't I," he whispers.

As My Father Retires

· I ·

My father's dreams are austere and, in an odd, inverted way, ambitious. In his lifetime, it is unlikely that he will achieve them, but he continues to announce his intentions, repeating them regularly, adding a detail or two, so that when and if the time comes, we will all recognize that he is a man whose future is no less than he had foretold for himself, had we only listened, had we only taken seriously what he said.

My father's dreams depend on my mother's dying before him, making his marriage a memory, more good than bad, a time when he fulfilled his obligations, was faithful, worked hard, and had the five of us, a long bridge between bachelorhood and becoming wifeless — a widower, the world will call him. My father's dreams depend on the expectation that Michael, Mary, Kelly, and I will allow him his eccentricities, not argue him out of anything or press our own ideas upon him. My father's dreams depend on the hope that we will leave him alone.

And that is why his dreams are, in an odd, inverted way, ambitious. We all assume, for one, that my mother will outlive him. Our family has been a family of widows, not widow-

he women — on both sides — living long past the men.

gh my father has added three years already to the age that has marked the limit for men in his line. At sixty-five, he has beaten the odds of all the fathers, uncles, cousins, and brothers who, like him, inherited the precarious combination of an Irish temperament, hypertension, and a poor heart. At sixty-five, he is already on overtime. At sixty-five, he is already old.

Second, if my mother *should* die ahead of him — for after all, after everything, the odds care little about lineage or expectation, the odds are really just one over the other — if he *should* become what the world calls a widower, it is highly unlikely that we will leave him alone.

And so, in a sense, it is to the air around us that my father forecasts his future, as we listen to him halfheartedly, unconvinced and unconcerned by what we hear again and again.

"After your mother goes, I'll be movin' on. I'll be makin' my own way," he tells us. "Take what you want before it's too late. After your mother goes, I won't be needin' all these extras," he says, referring to the furniture, the well-appointed living room, the curtains, the carpet, the centerpieces made of silk and straw. "Let's get rid of what we can while we can," he concludes. "Let's clear it all out."

Meanwhile, my mother continues to shop, to bring home bargains, to think in color schemes and collections, to build up barricades, bric-a-brac, against his better days.

All in all, my father is a man who does not own much — a house and its contents, a car, clothes that fill up one closet and a few dresser drawers. But even this is more than he wants to maintain. "Did you figure in the upkeep?" he asks each of us as we add one item or another to our lives — our own houses, our own cars, our own husbands and wives. "Everything is upkeep," he tells us. "If they say it's a dime, it'll cost you a dollar. Damn fools," he says as we follow in his footsteps, acquiring all the accoutrements of adulthood.

It is my father's dream to become happily dispossessed, to sell his house and everything in it, to fit what he needs into a footlocker and be on his way. Sometimes, being on his way means moving back to the city from the suburbs, renting a small semifurnished flat. Sometimes, it means nothing more specific than a place to sleep, shower, and shave. As far as I can tell, it never means a trailer, a camper, or life in a car. The only constant is the footlocker containing a few essentials. I don't know what is essential to my father. The footlocker is the only hard fact.

To me, the word *footlocker* has a military ring to it, but my father is not a military man. Like many men his age, he served in the army, but he didn't return to civilian life with stories or souvenirs, or if he did, he didn't save them to show his children or to tell when other men took out their stories and remembered the war. We were sorting socks one day. I was five or six. It was a Saturday, and my mother had begun working weekends as a receptionist at the medical center down the street. She left us a list that included folding the laundry. I didn't realize what these chores meant to my father: that he was stalled somehow, that we were in need of money, that he was looking for a way to move up. I was happy to be five or six and folding socks with him. We were in my parents' room, with the laundry basket on the bed. It was summer, I think. The room was sunny. My father didn't know our family's feet as well as I did. He kept pulling out his own socks and tossing the others to me. When he had finished all of his, he started unrolling the ones I had rolled up — mine, Mary's, Michael's, the few that were my mother's.

"Your socks are takin' up too much space," he said. "Look, look at mine."

I looked at his — brown, black, and navy blue — wound up tight like the roly-poly bugs we found under rocks in the back yard.

"Who taught you how to roll socks?" he asked, undoing another pair.

"Mom," I said. I didn't tell him that I thought it was a kind of magic trick she'd taught me, to hold two socks at the top and turn them in to each other, the two pieces contained as one. I had even developed a rhythm to it. I was proud of the way I rolled socks.

"Look," he said, unrolling a pair of pink knee socks. "You don't start at the top, you start at the toe." And he rolled the two pink legs together as if he were rolling a sleeping bag into a tight bundle, and then he turned the top of one sock over the top of the other to hold the pair together. Mine looked like shapeless lumps, like toadstools, next to his.

"Y'see," he said. "This way you can line 'em up in a row in your drawer," and he opened his sock drawer to show me the pairs lined up like troops of brown, black, and navy blue. It was the first privacy my father revealed to me, the first time he invited me to look at the way he arranged his life. I would return to look in this and other drawers many times on my own.

"How come Mom does 'em different?" I asked.

"Because she was never in the army."

"You were in the army?"

"I was in the Aleutians," he said.

I didn't know what the Aleutians were; I didn't ask. And it was not until several years later — in sixth-grade geography — that I realized my father had been referring to a place. "Aleutian Islands," it said on the list of places we had to find on the map. The moment I read it, my father's words came back to me. Until then, I thought the Aleutians were something he belonged to. "You were in the army?" "I was in the Aleutians," he had answered, and I heard it as if he were narrowing the field, telling me what part of the army he was in, which division or corps. The man next door was in the Army Corps of Engineers; my father was in the Aleutians. I had even boasted about it once. "My father was an Aleutian," I said to a boy down the street who claimed his father had been a high-ranking officer in a special unit of the army during the Second World War. "My father was an Aleutian," I told the boy, and he looked at me blankly.

Was it in the Aleutians that my father first became acquainted with a footlocker? Was it there that his plans for the future got started? I have never actually seen a footlocker, and even though I know it might be an ordinary item in someone else's life — just a trunk at the end of a bed — the footlocker in my father's future has become a kind of icon to me. Why a footlocker? I wondered. Why not a suitcase or a large safe deposit box? When my father's military life could be summed up in a sentence as short as "I was in the Aleutians," why a footlocker? Why not something from a longer segment of his life, something with which we were more familiar — why can't he just call it a trunk? — some kind of nondescript, neutral storage space that I could picture him filling with something other than rows and rows of socks.

"You're the in-tel-lec-tu-al," I imagine my father saying, slowly, emphasizing every syllable, in the half snide, half sincere way he sometimes speaks to me. "You were invited to study at the university — what would you suggest? What would be more in keeping with my circumstances? I'm not a man of words. I'm not one of the chosen. You choose. You probably want somethin' more symbolic. Somethin' with bricks in it, I bet. Take a man way back to his beginnings. Way back. Way, way back," he would say.

Something with bricks *would* be better. Something like the heavy white canvas carpenter's bag with brown leather handles that he keeps on the bottom shelf in the basement — the briefcase of a bricklayer who carried his tools from site to site. As kids, we pulled out the canvas bag one day — my father had gone from laying brick to selling it — to play with what we thought he had no interest in anymore.

"What is that?" I asked Michael when he took out my father's level.

"Lie down," he said. "I'll show you. Both of you lie down," he said to Mary and me.

We lay down next to each other on the cold concrete floor. Lying on cold concrete gave me my favorite feeling on a hot summer day. Michael placed the level lengthwise on Mary's stomach. "Don't breathe," he told her, and he knelt down to look at the little window of water. "You're about a bubble off," he said. "You're about a bubble off too," he said after he set the level on my stomach.

"What's that mean?" I asked.

"It means your bodies are off balance," he said, and he proceeded to test our outstretched arms and legs. Only our lower arms, left and right, were almost level.

"You probably have curvature of the spine," he announced after he made us lie on our stomachs so he could test our backs. Mine was worse than Mary's.

"Let me test you," Mary said.

"No," Michael said, refusing to lie down. "You can see with your eyes that I'm straight."

It was true. His body was as thin and indisputable as a perfect plumb line.

"Maybe the floor's not level," Mary argued. She refused to consider that she had curvature of the spine. "If the floor's not level, then we wouldn't be level either."

"Of course the floor's level," Michael said. "Dad wouldn't have bought this house if it wasn't level."

"Well, it's not," Mary said. She was lying on the floor next to the level.

"You don't know how to read it." Michael snatched the level up and carried it to another corner of the basement, where he set it down, stretched out alongside it, and saw for himself that it did not read true. He tested the floor in two more places. "Dad's gonna be mad as hell," he said, and he tucked the level under his arm and went upstairs.

My mother was in the back yard, working on her suntan. "Talk to your father about it when he comes home," she said, spritzing herself with water from a squirt bottle.

Michael, meanwhile, led us with the level from house to house until we had a band of friends following us. If someone's mother wouldn't let us troop into the house to measure the basement, we measured the carport instead. "Same difference," Michael said. By the time we got to the last house on the block, we knew that none of the houses in the neighborhood was level. Michael put the level behind his neck, hung his arms over it, and twisted his torso back and forth like Jack La Lanne on my mother's morning exercise show. We all circled around him.

"They'll have to bulldoze the whole block and build over. They'll probably put us up in a motel while they rebuild. That's what they do," Michael said, as we all walked back to our house to wait for my father to come home from work.

I was thinking about room service when my father pulled up. We were all standing on the side of the driveway, and when my father got out of the car, Michael delivered the news.

"Where's your mother?" my father asked.

"In the back yard," I said, thinking that he was as gallant as a man could be, to ask first, in such a grave situation, where he might find my mother.

"Well, let's all go inside and cool off," he said, taking his suit jacket from the hanger in the back seat. He never wore his suit jacket in the car during the summer. It was one of the things, he said, that increased his confidence, to have his jacket fresh when he made his calls. Later that same summer, in fact, in the style column of the morning *Globe,* he would be named best-dressed businessman in St. Louis, and as he pointed out his name to me in the paper, pinging his finger into the news-print, he said, "See? And I did that with only a few good suits."

When I was a child, I'd sit on the edge of my parents' bed and watch my father undress every day after work. He con-ducted his closet as if it were a second business, arranging his wingtip shoes, from brown to black, across the floor. They shone there under the darkness of his trousers, which hung by the cuff. I would watch him, standing in his shirttails, his gar-ters and socks, while he put away one day's pants, picked out another pair to wear the next morning, and hung them on the back of his closet door. His legs were as white as the moon when it's full, smooth as the silk ties he wore, and rich with blue lines of blood. I never understood, and still don't, why Michael, who sprawled out beside me on the bed sometimes, preferred to read from the evening paper, splaying our silence with facts, when each day he could have followed my father changing his clothes.

"C'mon, let's go in," he said, and he dispersed the small crowd of kids by telling them to come back to play after supper.

My mother was in the kitchen, and my father kissed her hello, hung up his jacket, and started down the basement steps, still not saying anything about the disturbing discovery that not one house on our street was level. We had assured our friends that my father would know how to handle the situation. After all, he had been a bricklayer, had built whole buildings with his bare hands. Now he sold the best brick in St. Louis, wearing the best suits a man could buy. He knew all the big contractors in town, Michael had boasted to calm down all the other kids, though no one really needed calming. We all accepted our fates rather casually. Maybe it was the heat. Maybe it was the fact that most motels had pools. Maybe it was that we were all past the age when we explored and exhausted the possibility of being adopted, and this new fact — that everyone's house had a secret flaw — gave us a communal opportunity to become outcasts, exiles, adventurers, to be set safely adrift under the leadership of our fathers. My father, I mean. It was his level, after all.

We followed him into the basement and watched as he picked up the rest of the tools we had left scattered on the concrete floor. After he put them back in his canvas bag, he turned to Michael.

"Can I have my level?" he said. "You can warp a level holdin' it behind your head like that." He put the level in the bag, closed it, carried it upstairs, and put it in the trunk of his car.

That was the pattern my father established for punishing us. Saying little or nothing, he would act quickly, without explanation, giving us a visual display of his disappointment, one that stung more than any words or reprimands, so that when he slammed the trunk of his car, with the tools locked tightly

inside, we knew that we could not be trusted, that private possessions had to be put out of our reach. He would use this
quick, silent action again and again — throwing the telephone
out the back door when we argued over it as teenagers; turning the hot water off in the house while we were still in the
shower, shampoo in our hair, if we didn't heed his warning
about water bills and shorter showers. "Turn it back on,"
we'd demand, dripping, shampoo sliding in snowbanks down
our shoulders, and he would shake his head and laugh.

But that day, no one was laughing. We had done what we'd
never done before. We had betrayed our father's trust, something that had existed quietly and without notice until we dis-
rupted the balance on which it was built. "A house settles" is
all he said when we sat down to supper. "Houses settle" is
what we matter-of-factly told our friends when we went back
out to play. It was the first time that something I didn't understand made perfect sense.

For me, then, the white canvas bag had more meaning than
a footlocker. "What about the white bag?" I could ask him.
"Your bricklayer's bag?" But I know he wouldn't find it appropriate. It would be like carrying his clothes in a pillowcase.
They would become wrinkled. He would look unkempt. My
father dreams of being a man unencumbered by accumulation,
but this — if I can count on anything from my father — he
will want to achieve without sacrificing his appearance. Perhaps that's the attraction of the footlocker: its shape and stur-
diness, the fact that it can't be overstuffed like a piece of soft-
sided luggage, its very rigidity a plea for order and propriety,
for the kind of packing that would preserve the pleats in my
father's pants.

"This is what they call the carry-on," he said, showing me
the piece of Calvin Klein luggage Michael had given him for
his birthday. He was fifty-five and packing for his first trip on
what he called a "commercial carrier." He had been on an air-
plane before, in the army, but that was a "far cry," he claimed,
from first class, a luxury provided by Michael.

My mother's matching carry-on was already closed and

standing in the corner. "She got that for no occasion. A free gift," my father pointed out, meaning that Michael had given it to her to complement his. I didn't tell him that I thought they were probably both free gifts, part of a campaign to sell Calvin Klein cologne.

"You're only taking these two pieces?" I asked.

"It's all we're allowed."

"It's all you're allowed to carry on. But you can check larger suitcases at the baggage counter. Two apiece, I think."

"No, your brother says it's best we stick to this."

"Why?"

"Oh, y'know your brother," he said, placing four pairs of underwear on top of his pajamas. "He gets impatient with the processes in life."

"But he's not going with you."

"Well," he said, "I guess he thinks we'll be befuddled at the other end."

"They do speak English in Boston."

"Well, your mother gets flustered. Y'know."

I stretched out on the bed next to his suitcase and thought of my mother saying, "You know how nervous your father gets." The ceiling fan circled on slow above us, but it was not summer. It was fall. In any season other than winter my mother is warm. The fans that my father installed in every room — he had bought them on sale, in several styles — were a reminder, left running, of wherever she had been. I listened to the fan; a slight pause in each rotation made its movement sound irregular, and it was that pause, that imperfection, that made the room feel familiar and safe.

I leaned on my elbow and watched my father pack. He took two handkerchiefs out of his drawer. We used to buy them ten for a dollar — something as kids we could always afford to give him for Christmas. The expensive ones, sporting his initials, would come from someone else — my mother, my grandmother, or my aunts. When he was a salesman, Mary and I used to iron them perfectly, for ten cents apiece. We pressed the monogrammed ones so the TJF would poke out of his breast pocket at just the right point, when that was the style. "High maintenance on these handkerchiefs," he used to say, handing me a dollar. "You girls got a real racket goin'. Pay out a buck for a present, make it back a hundred times over before the next holiday."

"Do you want me to iron these for you?" I asked, picking up the handkerchiefs from his suitcase.

"Can't afford it," he answered.

"I'll bill you," I said, but he had already gone on to flattening his shirts.

I can't remember exactly when we stopped ironing my father's handkerchiefs. When I try to remember, I have the sensation of a shadow moving over our lives, the feeling that I missed a particular moment, much like the day I slept through the total eclipse of the sun.

"You missed it," Michael said when I woke up. The moon had moved on, but he was still standing in the front yard with the cardboard viewing box on his head, its pinhole pointing to the sun, as if he had been an integral part of the alignment.

"What did it look like?" I asked. I hadn't understood the whole week's flurry of preparation for the total eclipse. Unlike every other kid on the block, I hadn't built a viewing box. I was seven, and I didn't believe the warning about looking directly at the sun. Why would it be more dangerous to look directly at the sun during an eclipse? I kept asking. Wouldn't that be the best time to look directly at the sun — when it was blocked by the moon? When the moon in effect masked the sun's usual strength? Just don't do it, is all anyone would answer. But every day, while the other kids were constructing their boxes so they could look at the solar eclipse safely, seeing it projected on a piece of paper like a peep show, I dared the sun by squinting at it directly. Except for seeing dark spots in front of my eyes for a few seconds, I suffered no side effects. As it happened, I fell asleep the afternoon of the eclipse, and when I woke up, it was over. All the kids on the block except Michael had abandoned their boxes and gone on to something else. The sun was at its full strength in the sky; the moon, a moment past it, looked self-assured to me, and satisfied.

I knocked on the side of Michael's box. "Why are you still standing here?" I asked.

"I'm trying to figure out what went wrong."

"What do you mean? Didn't you see it?"

"I saw something," he said, "but I thought it would be better."

And that is how I think of my father's life, as if there were some moment we were all expecting, some phenomenon in the future for which we all helped to prepare — ironing his handkerchiefs, pressing his shirts — some success, blindingly bright, that was sure to be his. But before it ever came, it was suddenly over. Who would have thought the brick business would bottom out, that whole houses, whole subdivisions even, would be built without brick, that there would be so little left to sell. And like Michael, my father thought that if he had only built a better box, things would have been different. And it was that disappointment, that despair, that we could not bring ourselves to look upon directly.

"What are you going to pack here?" I asked, pointing to a corner of the carry-on that he was obviously avoiding.

"I'm to reserve that space for your mother's makeup," he said.

"Are you sure you don't want to take a second suitcase?"

"Well, I don't wanna get your mother started. Give her any ideas. She'd have to make a whole new set of decisions about what to bring with her. Fact is," he said, putting his bottles of prescription heart pills into a plastic bag, "my father went all the way to Saudi Arabia and back with a suitcase smaller than this."

I can't say for sure why my grandfather went to Saudi Arabia and back with a suitcase smaller than a Calvin Klein carry-on. "He was in the Foreign Legion," my father told me the one time I asked, as if, like "in the Aleutians," it was a self-sufficient fact. I was nine — almost ten — by then. Sean had been born the year before. We were living in our second house, a two-story, all brick, which had been built according to my parents' specifications on the last remaining lot of a subdivision that had been developed eight years earlier.

"American Heritage Red," my father told me, rubbing his hands on the brick as we walked around the house. "Smoke-gray mortar," he said, running his index finger along the line of hard mixture between the bricks. "It's a good combination, don't you think? I came up with it myself. I've looked at a lot of houses over the years. I've seen some mistakes, where the brick and the mortar were of just no benefit to each other. The owners probably never knew any different. But we got it right. American Heritage Red with smoke-gray mortar. If you stand back, you'll see. It's subtle." But we couldn't stand back, because the yard had just been seeded, and for some reason we never took the time to stand back later, after the grass had grown in and the house was no longer new.

"Do you think a house could be built without mortar?" my father had asked me that day.

"Doesn't look like it could."

"I think it could. I have an idea," he said as we turned the corner to the back of the house. He put his hand on the downspout of the gutter and tried to move it back and forth with a motion like the one he used to straighten his tie. "Do you wanna hear?" And he took my hand and walked on, the downspout having passed his inspection.

I looked up at the windows of my parents' room — the master bedroom, my mother called it. White shades with scalloped edges decorated every window on the second story, and each shade was pulled down to exactly the same position, forming an even edge around the top of the house. The windows were dark, more narrow than wide, and the row of them, with their white scalloped shades, reminded me of a photograph I'd seen on the society page of the *Post-Dispatch* the previous summer. Five women stood in a receiving line, all wearing dark sleeveless gowns and long white gloves that stopped at exactly the same place above their elbows. That is how our windows were, under my mother's direction, each shade pulled to the same polite position, reserved but welcoming. As I looked up that day, my hand in my father's, I imagined my mother describing the interior details of the house to Mary, and I wondered whether my parents knew instinctively the kind of life each of us would lead and what we would need to know to lead it.

"My idea," my father said, "is interlocking bricks. Bricks that fit together like puzzle pieces. It would change the market, do away with mortar altogether. The interlocking part is simple. The hard part is figurin' out how to reinforce the structure."

I was confused and disappointed but didn't say so. It was all my father ever said to me about his idea, and though he presented it to me seriously, I never saw any evidence that he was serious about it. There were never any strangely shaped bricks in the basement, no scale models or sketches. Why would a man want to do away with a material that meant so much to him, I wondered, a material whose composition and qualities he understood? On one side of the house, he spoke of subtlety. On the other side, of making a change in the market, one that would make mortar obsolete. And it was not just the material that would disappear, but men's lives — bricklayers, tuckpointers, the whole tradition of apprenticeship of which he had been a part. And what about the aesthetics? What about the beauty of pairing red brick — "It has a touch of blue in it from the way it's fired" — with smoke-gray mortar, or the clean line that's created, the solid, sturdy assurance of predictable patterns, of rectangles in a row? How long could you look at a house with the irregular wave of a jigsaw running through it? How long could you live in it? That was my confusion. That was my disappointment. And this, too: I thought my father's attitude toward mortar — admiring it one moment, doing away with it the next — meant that he loved things in a disposable way. And at a time when the world was becoming a more permanent place to me — we had this house, after all, and we had Sean, who had added longevity to our lives, enlarging our family after it had stayed the same size for nine years. Sean was the first person we had all started to love at the same time, none of us, except maybe my mother, knowing him ahead of the others. When we had all this to hold on to, things that I was sure would never end, my father introduced me to the idea of impermanence.

At the same time, I felt privileged that my father had entrusted me with his idea, but I wanted the idea to be one I was incapable of understanding, not because I wanted it to be better, but because I didn't want the way he talked to me to change, and that day his words seemed different. I wanted him to remain a man who said things that sounded enigmatic, things like "a house settles," which I could accept only on faith and my father's authority. And though I didn't think immediately of the objections I had to his idea, nor at nine could I have articulated my confusion or the consequences of a change in the market as I would later lay them out, I was struck by the realization that someday I would fail my father, that I would betray him, if only by growing up. And I felt a sadness as I stood there, the brick wall behind us, ahead of us a yard of seed and straw.

It was only a few days later that I found the suitcase my grandfather had carried to Saudi Arabia and back. When we moved to our new house, things surfaced that I had never seen. I was lying on the basement floor racing my roller skates. I pushed both skates from a starting line to see which one rolled farther. My left foot always came in first. I stood a Barbie doll in each skate — my blond one and Mary's brunette — and sent them off, each one waving from the top of a white shoe skate like floats in a parade. I pushed them faster and faster, until one skate went out of control and tipped over, shooting the brunette Barbie beneath a set of shelves my father had built. There was just enough space under the bottom shelf for total darkness and a Barbie doll. I stuck my arm under and felt around, but I couldn't reach her. When I tried to reach farther, stretching hard and straining my shoulder, my head was at an angle that put my eyes even with everything on the bottom shelf, and there, behind a cardboard box marked GOODWILL, was a tan leather suitcase I had never seen in our old house.

I left Mary's Barbie beneath the shelf and pulled out the suitcase. It reminded me of the suitcases the elephants carried on their backs in the Saturday afternoon safari movies Michael and I watched on TV. Two brown leather straps were wrapped around it, and a sticker showing a flag with a sun was peeling off the side.

The suitcase smelled stale, and opening it was like staging a small scientific experiment in which molecules of modern air mixed with those from a much earlier era. Inside were newspaper clippings, labor union medals, and religious mementos. The clippings were from the *Labor Tribune*. One announced my father's apprenticeship, slated to begin under his father the following week at the Granite City Steel Mill, where bricklayers worked twenty-four hours a day relining the inner walls of the coke ovens as they cooled. ("It was like workin' in hell," my father told me later. "Y'had to wear thick gloves when you laid the brick or your knuckles would just burn out bare.") Another article announced his father's death. It included a picture of him and words about his union loyalty and leadership. He looked like a laborer, like one of the men who come up from the mines, though it was not dirt that defined him. He was as clean as a man could be, but his eyes were like the eyes of those men just up from the mines, filled with a dry fatigue, deadened but still determined, a gaze that showed neither rage nor resignation. He looked nothing like my father and nothing like me. He had died a year or two before my parents met, just after my father finished his apprenticeship, and to me, the fact that my mother never knew him took away from the legitimacy of his life, for although my father was a better storyteller, my mother was the more accurate source for who people were and how they behaved in the world before we were born. In an unsealed envelope were the obituaries of both my father's parents. What I know of my grandmother, who died before I reached an age to remember her, was that she supplemented my grandfather's earnings by selling homemade egg noodles to her neighbors and writing radio jingles that never aired at a time when anyone was awake to hear them. I had seen other pictures of her. She was small and looked opinionated.

A cardboard box held a couple of union pins commemorating my grandfather's years of service and a large silver ring with a pyramid and palm trees engraved on its surface. I pressed the ring into my forearm, and the image appeared on my flesh like an inkless tattoo. I stamped a row of pyramids and palm trees from my elbow to my wrist, but by the time I finished, the first had already faded. I got a stamp pad from my father's desk, pressed the ring against the ink, and applied it just above my ankle. All I got was a solid black smear. I inked the ring again and wiped it gently so that only the etched lines were dark against the silver. This time I made an imprint in the center of my palm that was so perfect it wasn't worth repeating. I spit on the ring, wiped it clean, and put it back in the box.

All that was left were the religious items. There was a prayer book published by the Confraternity of the Precious Blood, a rosary with brown beads, a few holy cards, and a locket unlike any I had ever seen. When I opened it, a tiny statue of the Virgin Mary fell into my hand, adding a third dimension to the scene I had stamped there of the pyramid and the palm trees. I opened the locket several times to make the Virgin Mary appear and reappear, pretending at first that I was one of the children at Fatima, and then, tiring of that, moving on to Lourdes, and playing out the scenarios of all the shrines I had learned about in school. Finally, I hung the locket around my neck and tried to figure out how I could keep it. Having hit on a plan, I packed up everything and went upstairs to find my father.

He was in the family room reading the paper. "Dad," I said, "I lost a Barbie in the basement."

"What do you mean?" he asked, without looking up.

"She's under those shelves and I can't reach her."

"How'd she get under there?"

"She was thrown from a float in the St. Patrick's Day parade," I said. "She's probably dead, but I need to get her out anyway, because she's not mine, she's Mary's."

"Sounds serious," he said, still not looking up.

"It's very serious. I told Mary I wouldn't play with her stuff when she's not home."

"Then why did you?"

"Because she's not home."

"Okay, c'mon," he finally said, putting down the paper. "I'll bail y'out."

In the basement, my father stretched out in front of the shelves and felt around for Mary's Barbie.

"Did you see where she went under?" he asked.

"Right about where you are," I said. I lay down behind him and put my head on his hip.

"I don't feel her," he said. "Maybe we need a flashlight. Oh, wait, here she is," he said, and dragged her out by the ankles.

"Hey, Dad," I said, my head still on his hip, "what's that?"

"What?"

"Behind that box."

"Oh, that's my father's suitcase from Saudi Arabia."

"Why did your father go to Saudi Arabia?"

"He was in the Foreign Legion."

"Is it empty?"

"There's just a buncha old junk in it," he said.

"Can we look?"

"It's nothin' you'd be interested in."

"Let me see," I said. "I've never seen junk from Saudi Arabia."

"It's not junk from Saudi Arabia. It's just a few mementos of mine from my mother and father."

"Let me see," I said. "Please."

He took out the suitcase and put it on the Ping-Pong table. The air the suitcase released was less stale than when I'd opened it an hour earlier, and I looked at my father to see if that made him suspicious.

"See, it's nothin' much." He picked up one of the pieces of newsprint. "That's my father."

"Handsome," I said.

He shuffled through the holy cards and set them aside.

"What's in here?" I asked, pointing to the cardboard box.

"Pins that you get from the union after workin' so many years." He took one out for twenty-five years and showed it to me. "And this ring," he said, putting it on his finger, "was somethin' my father bought in the Middle East."

"When he was in the Foreign Legion?"

"Yeah," he said.

"Why don't you wear it?"

"Oh, it would upset your mother. I don't wear a weddin' ring, y'know, because when your mother and I got married I was layin' brick for a livin' and a ring would just get all banged up. I don't think your mother ever quite believed that, so if I wore another ring now, it'd probably hurt her feelings. Besides, it's not my style," he said, and dropped it back in the box.

"Whose rosary?" I asked.

"My dad's." He picked it up and let it fall through his fingers. "See, not much," he said. "Your aunts and your Uncle Bud got most of it. Let 'em keep it. I didn't enter into it. I wonder what you kids'll claim when I'm gone."

"What's that?" I said, pointing to the locket.

He picked it up and let it dangle from its chain.

"You think that's pretty?" he asked.

"Sort of."

"Well, there's more," he said. He spilled the statue into his hand and held the Virgin Mary between his thumb and index finger for me to see.

"Wow! Can I try it on?"

"I guess," he said. He put the statue back in the locket, undid the chain, and fastened it around my neck. "That belonged to my mother, but I never saw her wear it."

"I'd wear it all the time if it was mine."

"That'd be a sure sign you'd lose it," he said, and he un-clasped the chain and removed it.

"I wouldn't lose it."

"You'd forget you were wearin' it and start roughin' around."

"No, I wouldn't."

"I know you," he said.

"I promise I wouldn't."

"You really want it?"

"Yeah."

"Okay, then," he said, without saying anything about shar-ing it with Mary. That was the difference between my mother and my father. My father didn't see the need to treat us equally. My mother balanced us like a budget. Both ways worked to our advantage, but my mother said that the burden always fell on her to settle the fights between us that my father caused.

"Thanks," I said as my father placed it in my hand.

"What's this?" he asked. I had opened my hand wide, for-getting about the pyramid and the palm trees.

"Oh, just a stupid fake tattoo from the dime store." I pulled my hand away.

"Let me see," he said. He stared at the pyramid and the palm trees, then closed my hand around the locket and pushed it away. To this day, I don't know whether he recognized the imprint as the engraving on his father's ring or, if he did, why he chose not to say anything about it. It was worse not know-ing, and the locket was like a lump in my throat. I didn't even show it to Mary, and a few weeks later I lost it. Since my father never followed up on the things he gave us, I didn't have to confess to him that the memento of his mother ended with me. I had no idea where I lost the locket, but for a long while I worried that he might come across it someday while he was mowing the lawn, a glint of silver smashed into the soil that he would bend down and dig out with his finger. And I wondered, if he ever did happen to find it, whether he would hold on to it or give it back to me, whether he would wash it off or just place it on my dresser, dirty and dull.

Going through my grandfather's suitcase from Saudi Arabia was one of the few times my father ever told me anything about his parents. Now, years later, packing for his trip to Boston, he offered another fact.

"My father was born a twin," he said. Maybe it was because he was heading east, where his father was from — New York, not Boston — that made him think of it. "His twin brother died shortly after birth. Thing is," he said, "they got the names mixed up. My father was really the one named Patrick Francis. It was the one named Martin Joseph who died. My father went on for the rest of his life with the wrong name."

"I don't get it," I said.

"What's not to get?"

"Well, how'd they find out?"

"Someone recognized the error, I suppose."

"Well, who would? I mean, they didn't do fingerprints and footprints and all that back then, did they? And if they were identical, really, who'd know? Did your father have some kind of distinguishing mark that made them realize the mistake? If they realized, they must have realized pretty soon after the fact, so why didn't they just start calling him by the right name?"

"Goddamn it, it's just a simple story. You miss the point."

"What's the point?"

"Point is, I can't tell you kids anything. Can't pass on a piece of information or interesting fact that you don't try to contest it in some way."

He finished packing his Calvin Klein carry-on. "Why did you decide to go to Boston?" I asked.

"Well, your brother's been there. He recommended it. When you can go anywhere and you've never been anywhere, it's kind of hard to decide."

"Quite a birthday present, huh?" I said. "A weekend anywhere you want."

"Funny, isn't it, that TWA lets employees' parents fly for next to nothin'. Doesn't sound very profitable."

"Well, don't worry about it. They'll probably stay in business long enough for you to get to Boston and back."

"I'm not worried about it."

"What are you worried about, then?"

"I'm not worried about anything."

"You look worried."

"I'm worried I didn't leave enough room for your mother's makeup, that I'll have to updump all this and start over."

"You left enough."

"I'm worried you're blockin' me in."

"I'll move my car before morning. Anyway, Michael's driving you to the airport."

"I know, but I hate bein' blocked in."

"Okay, I'll move it now."

"You don't need to move it now. What time do you plan to retire?"

"I don't know. When I get tired."

"And what time are you scheduled to rise?"

"I don't know, Dad. When I wake up."

I was staying over to take care of Kelly, who was still in grade school. When my parents left the next morning, it would be the first trip they had taken together, alone, since the five of us were born. Michael had arranged the weekend, proposing it as a birthday present for my father. He had suggested it two weeks earlier, when we'd all been together for Sean's sixteenth birthday. My mother made Sean's favorite meal, lasagna with garlic bread, and we ate in the dining room using the good china. It was the way we had celebrated his birthday every year, and it seemed natural that we do it even though he was dead. Well, maybe it didn't seem natural. No, there wasn't anything natural about it. It felt necessary, like something we desperately needed to do, to have a day that was openly and outwardly about him, one in which we didn't have to worry about where we were with our grief. From that year on, it would be different. There would be no more planned celebrations. At most, my parents would have a Mass said for him every year on October 2, and individually we might commemorate the day by saying, "It's Sean's birthday," to ourselves or to each other, thinking about how old he would have been, wondering what he might have looked like.

But that year, the first year, we were all together on the day Sean would have turned sixteen. "I would've liked to have seen him drive, for him to have driven a car, y'know, for him to have . . ." My father's voice trailed off, and he started to cry. It was then that Michael suggested the trip.

For Michael, travel was an antidote for everything. He had been working for TWA for a few years by then, but he had never been able to persuade my mother and father to take advantage of the flight privileges available to parents. This time they consented. They would take a short trip.

My father put his carry-on next to my mother's in the corner of the bedroom. Behind the suitcases a glass jar filled with nickels, dimes, and quarters stood as high as my father's knee. I wondered what he'd do with that money now that Sean was gone. They had been saving their change to take a trip. I lay on my parents' bed while my father went into the bathroom to put on his pajamas and robe. When he came out, he went downstairs to say goodnight. He would first kiss Kelly, holding his robe closed while he bent over, and then my mother, saying, "I'm goin' on up." If I were down there, he would ask me again, before kissing me goodnight, "At what time do you plan to retire?" And then, "At what time are you scheduled to rise?" Often, his formality began as a way of being funny. Usually, it went on to become a routine. Sometimes, I made up an answer to appease him. It worried him that I stayed up until all hours on my nights off and never got up before noon. "You have to discipline yourself the same as if you were on the day shift," he'd tell me, a disciple himself of going to bed every night at nine-thirty and waking up each morning at a quarter to five.

"Get up," he said, "I'm goin' to bed." He put his glasses on the dresser, hung his robe on a hook behind the bathroom door, and dropped his dentures in a cup on the sink. ("When did you get dentures?" I asked him, startled, the first time I saw him take them out. "Back when you were busy with your own life," he answered.) I started to leave the room. The ceiling fan was still rotating slowly above us. The light was out, and I heard my father pull down the bedspread and the covers and lie down. The next night he would be sleeping in Boston. If I mentioned this, he would say it was no big trick for a man his age. And he was right. His big trick had been bringing home a paycheck every week since he'd been married and turning it over to my mother, who managed, even in the most meager times, to make it do. His big trick had been hanging on in the years between the end of the brick business and the beginning of his life at the Bulk Mail Center, where he sorted packages as they came off a conveyor belt, the years between ambition and bouncing back.

The first year of the years in between he bought a *Post-Dispatch* newspaper route that had been losing subscriptions, but despite his hard work, he couldn't revive it. He bought a secondhand van and a string machine, but he couldn't afford to hire anyone to help him, so he set out rolling the papers himself, passing them through the string machine, and tossing them out the window as he drove by each customer's house. If it was raining or if the prediction for rain was greater than 50 percent, he had to wrap the papers in plastic. Sometimes Michael, Mary, and I would go with him, rolling the papers and working the string machine while he drove and threw.

"Roll 'em up a little tighter," he would tell us, "so they'll take less string." The string machine was like a third arm in my father's operation. At the front of the machine, a string was threaded between two points, and each time we tapped a newspaper against it, part of the machine would bow down, loop the string around the paper, knot it, cut it, and advance another length of string from a spool. To me, the machine seemed like a woman, graceful, efficient, and unfailing — tying string, cutting string, pulling string — with a sound that set our pace. If we were steady, the sound was constant and approving, a kind of mechanical praise. If we were slow and erratic, the sound was disjointed and unsettling. So we worked to win the rhythm, and it made me think there was something maternal about the sound of machinery that kept men laboring all their lives. And if my father wished for us to have a better life than the one called labor, in my case the rhythm worked against him. I was seduced by the sound, and I knew that if I never got an education, I would be more than satisfied with a skill.

In the evening customers would call and complain. They hadn't received their paper. Or it had rained unexpectedly, leaving the paper wet and unreadable. Could he run over with another? Why not? Could he figure in a refund? Of course. Could he deliver a few weeks' worth for free? a woman asked. Her husband was out of work and needed the want ads.

In those days, my father kept bags of loose tobacco in his desk drawer and a hand-held machine for rolling cigarettes. He couldn't afford the Camels he usually smoked. Every evening he would go down to his desk in the basement and roll the twenty cigarettes he allowed himself each day.

"Show me how," I said.

"The trick is to pack the tobacco tight," he told me, demonstrating how to tamp it into the barrel of the machine. "Then you curl the paper around this track, lick the glued edge very lightly, and send it on through." Sending it on through meant releasing a lever that shot the cigarette paper through the middle of the machine, where it was filled with tobacco, sealed, and sent out the other side. The cigarettes looked short and unsophisticated, even compared to Camels, which my father said were the truest type of cigarettes you could buy, not "fancied up or filtered."

"Let me try," I said, and just as he had taught me how to roll socks and how to roll newspapers, my father taught me how to roll cigarettes, unrolling the loose ones and making me roll them over until I had done twenty that were tight enough to set side by side in his silver case.

After a year, my father sold his secondhand van, his string machine, and his route of dwindling subscriptions and went back to laying brick inside the coke ovens of the Granite City

Steel Mill. Twenty-four years had passed since he had apprenticed with his father. He was no longer a young man; he was a middle-aged man who fell ill and went on working. A decade later, doctors would determine that he had probably had several undetected heart attacks when he was in his forties, making him a man with a history of heart trouble. If it weren't for the five free points that the federal government tacked on to the test scores of veterans taking the post office exam, which enabled my father to leave the steel mill for a job sorting mail, he might not have lived longer than most of the men in his family, whose lives seemed destined to end shortly after they reached the age of sixty.

I looked at my father lying in bed. The light from the hallway illuminated his head and his arm. On the dresser, his glasses gave off a glow. The closet door cast a shadow on the carry-ons in the corner. Behind them, the jar of quarters, dimes, and nickels shone softly, as if the coins had been minted with moonlight.

"Try to persuade your mother to come up at a reasonable hour," he said. "Your brother'll be here early."

"I'll try."

"Maybe we'll go on a trip sometime, just you and me, after I retire," he said as I lingered in the doorway. "We move at the same speed. We'd be well suited."

"Maybe," I said, and a few nights later, as I watched him unpack, he said, "You'd like Boston. Lotta old brick."

Each evening, as my father retires, he turns on the radio, tuned to a talk show, and lets the sound put him to sleep. "This is Jim White at your service," the announcer says as he answers each call. "You're on the air, caller. Where are you calling from?" "Belleville. Red Bud. South St. Louis," they say. As a child I would fall asleep next to Michael, Mary, or my father — depending on the order in which we joined him in bed — imagining a giant map on my parents' ceiling, constellations of cities lighting up with each call. "Ferguson. Florissant. Maryland Heights."

"Recognize that guy?" my father would ask. "That's Gino down the street," he would say from his side of the bed, until it became a joke in the dark. Any caller my father referred to as a "fast talker" or a "slick operator" was "Gino down the street."

"Is that Gino down the street, Dad?" Mary would ask.

"That sounds like Gino down the street," Michael would say.

"What side of the street does he live on?" I'd ask.

"You're so dumb," Michael would say. "He doesn't really live down the street."

"So why does Dad call him Gino down the street?"

"It's a joke, Stinky," my father would say. "Now go to sleep."

"This is Jim White at your service," I say to myself now, alone, on nights when nothing else works. In my mind, my legs are the longest, and when we all hold them in the air, I win. "What are you doin'?" my father asks without opening his eyes. He's lying on his back, shirtless, beside me, one arm under his head on the pillow, the other bent at the elbow like a salute, his hand over his eyes as if he is shielding them from sudden light. Each night, he lies narrow and still, as if he is sleeping not in a bed but on a park bench, ready to wake up right away. "What are you doin'?" he asks again. My legs are down now. My arm is up. "I'm counting the cities on the ceiling," I say. "There are no cities up there besides St. Louis. Now go to sleep," he says, pulling my arm down in the dark.

When I was older, I learned that it was my mother, not my father, who moved us into our own beds in the middle of the night, a task she later turned over to me. "If you're going up, could you put the kids in their own beds?" she'd say, reading her *Redbook* as she lay on the couch. Upstairs, Sean and Kelly would be asleep on their backs beside my father, the three of them so close together there was room on the bed for one more. Until my mother came up after midnight, the space would go unoccupied. In some ways, what my mother said was true: Michael, Mary, and I were their first family, Sean and

Kelly their second. I would look at my father lying in bed with his second set of children and wonder whether he was the same father to them as he was to us, or whether the years between ambition and bouncing back, or even the simple fact that he was older, made his relationship with them different.

On those nights when my mother asked me, I would carry Sean to the bed he shared with Michael. Putting his head on the pillow, I'd kiss him on the lips, causing him to grind his teeth a little, and he'd turn on his side, his back to where Michael would later lie. He was as undisturbed by these nighttime transitions as I had been, content to fall asleep in one place and wake up in another. A toddler, Kelly traveled with her own blanket. When we stopped in the bathroom, I held one end of the blanket while she pressed the other to her cheek as she sat on the toilet asleep. After I laid her in her bed, kissed her, and covered her, I'd look back in at my father, lying in the same narrow space, the rest of the bed now empty. "This is Jim White at your service," the radio would say. Once, as I stood in the doorway and listened, a woman called in, filling the room with a voice that was immediately familiar. She spoke insistently about mistakes the mayor had made with the unions, how it would show the strength of labor not to reelect him. As Jim White tried to get a word in, my father sat up in bed, turned to me, and laughed, as if he had known all along that I was there. "Did you hear that?" he said. "Even Jim White can't contain her," and he lay back down and listened to the rest of what his sister had to say from South St. Louis.

From what my mother tells me, I imagine my father sprinting out of bed from his park-bench position on the night Sean died. For years, he had been ready to wake up right away, and when he heard my mother screaming his name up the stairs, he ran down and found Sean collapsing at her feet in the front hallway. She had already called 911. My father carried Sean, barely conscious, up to his bedroom. The floor, the walls, the bed were covered with vomit. My father found a clean place to lay Sean and stripped him of his clothes. They were soiled with sickness. He took a fresh shirt from Sean's closet, underwear and jeans from his drawer, dressed him, and carried him back downstairs. Why did he do this? Why was it crucial to him that his son be dressed in clean clothing? My mother waited in the hallway while my father ran back up and put on his own clothes over his pajamas. Only the three of them were home: my mother, my father, and Sean. My father left the radio on. The ambulance arrived.

A few hours later, my father walked into his bedroom. Jim White at your service was off the air, and the radio emitted only static. Would I ever tell him about the time, two months earlier, when I had picked up Sean late one night from a party? After we'd dropped off all his friends, he turned on the radio and tuned in to Jim White. We listened together to the people calling in. A man pitching burial plots for pets phoned in to talk about the proper way to bury a dog. Sean looked at me and grinned. "I think that's Gino down the street," he said. And I knew then that we were one family, not two. In more than our blood, our father belonged to all of us, one the same as the other. In more than our blood, we each belonged to him.

"I'm in 2E now," my father tells me when I stay over one weekend. He is talking about the bedroom that used to be mine and Mary's, on the east side of the second story. "I'm goin' on up to 2E now," he tells us as he kisses each of us goodnight, holding his robe closed as he bends over. My father is in 2E now because my parents have branched out in the house, taking their ill-matched sleeping habits to separate bedrooms. It is going on ten years since Sean died. At night, my father screams in his sleep, starting with a low moan that escalates in seconds. I lie in my old room and listen. Kelly wakes up and runs in. She is a nursing student now, and I tell myself that she knows better than I what to do, as if his bad dreams have a medical basis. She is willing and she runs to him. Most nights, on the other side of the house, my mother sleeps too heavily to hear his screams, though when she does, she is the one who wakes him.

"How long has he been doing that?" I ask my mother the next morning.

"He says he never dreams about him," she answers, responding to the question she knows I will not ask.

Does he dream about me? Does he dream about Michael, Mary, Kelly, and my mother? Does he recognize us as we appear at different ages? In his dreams, does he call each of us by the wrong name? Does he have hopes for any of us beyond the hope that we will be happy? "I just want you all to be happy" was the only condition he put on our futures. He and my mother shared the same view. They had faith that we would each find our way. Without being able to offer us an apprenticeship in anything, they taught us what they knew. They taught us how to work. And thinking that happiness somehow let us off the hook, I, for one, rambled around a while with no real ambition, ruining one opportunity after another, thinking the mistakes I made would have no effect on the future if in the end I could say I was happy.

· 2 ·

Freed from happiness, we could create any future for Sean that we wanted.

"What do you think he would have become?" my father asked me. I was raising the seat on Sean's bike so my father could ride it.

"I don't know," I said, though I was thinking a veterinarian or some kind of scientist. Veterinarian was what he sometimes told people, and then pediatrician popped into my mind as I was pulling up the seat. In the four years that had passed since Sean died, the bike had been hanging from the garage rafters, unridden. "I have no idea," I said. I could be mean that way when I wanted, putting a quick end to his questions, not looking up to see what his eyes had to say. Almost immediately, I felt bad about it.

Older. He would have become older, I wanted to answer. He would be nineteen now, the age when for me real failure started. I had flunked out of Northwestern, losing a full scholarship, leaving before the year ended. My father had come up by himself to drive me home. It was a Tuesday in the middle of March. "Goddamn it, it's my only day off," he said, pounding the steering wheel after we had driven over a hundred miles without talking. Sitting beside him in silence, I thought of all the letters he had sent me while I was away, each of them written in the early morning, when he sat at the kitchen table alone with his coffee before leaving for work. "Your mother, Sean, and Kelly are upstairs asleep," he always started, "and I have some time to spare before punching the clock." Each time I read that line, I understood it as my father's way of saying that everything was right with the world, everyone under his charge that day safe and sleeping above him. He would be gone before they got up. I was surprised when the first letter arrived. It came directly from him — addressed by him, mailed by him — not merely a note attached to the letter my mother sent every week. "Blew you round the bend that I could write, did it?" he had asked the next time we spoke on the phone. And yes, I had to admit that it did, but only because he had always enlisted my help in writing letters when he was unemployed — cover letters and queries — swearing, damn it all to hell, that he was worthless when it came to words. We would sit at the kitchen table, and he would tell me what to write. "But smooth it out," he'd say. "Make it sound good. Know what I mean?" And so, yes, I was surprised. He could have done it alone all along.

"I guess y'know what you're doin'," my father said when we were nearly home that day, my college career having ended. I wanted to answer, but I didn't know what I was doing. I was failing, and I wasn't sure why. I couldn't concentrate. I couldn't focus on anything long enough to finish it. This had started in high school, but I was able to cover it up. In college, I couldn't. I kept reading the same sentences over and over, and they didn't make sense. It would be several years before I was diagnosed with severe depression. Until then, I struggled, redefining success in terms that were less and less ambitious. I would begin my adulthood as a carhop at the

Steak 'n Shake up the street, move on to become an assistant in a photo lab, a mail carrier, a proofreader, a printer, and, finally, a writing tutor at a community college, where I had returned to school part-time. What would Sean have become? What would I become? I wondered, tightening the bolts on his bicycle seat so my father could ride it around the block — give it a spin, he said, see if bike riding was the kind of exercise he was after. I had tried to convince him that one of the other bikes might be better — Sean's touring bike or one of the ten-speeds I'd left at my parents', all of them hanging in a row from the rafters. But no, he wanted to ride the racer, Sean's good bike, with its thin tires and featherweight frame. Knowing nothing about gears, he had asked me to come over. Before showing him how to shift from one speed to another, I pulled up the seat, raised the handlebars, and removed the toe clips, imagining Sean doing the same. Then be kinder, Sean would have told me. Yes, be kinder, I thought. Veterinarian, or some kind of scientist, I wanted to tell my father, but no words would come out. He stared at me as I fooled with the bike, willing me to look up, but I kept my head down and went on working. Was it he who had run behind me, teaching me to ride one?

I finished explaining the gears, and he straddled Sean's bike, ready to go. "Tom, your heart," my mother would have said had she been home. It was what she said whenever he exerted himself. It was the sum of him now, the whole of him: his heart.

I watched him coast down the driveway and turn on to the street, wobbling a bit. Then, without looking back, he steadied himself, went down the hill, and disappeared around the corner.

I sat down on the driveway and waited. It was a Tuesday, of course, my father's day off. He had Mondays off too, if he didn't take the overtime, though it was a rare day that he'd refuse it. Down the street, he returned from around the corner and rode in slow circles at the bottom of the hill, getting used to the gears, and then he took off again. His non-scheduled day. That's what he called it, using the language of the mail-sorting plant where he worked, pronouncing it "*nun*-sched-uled" and making me wonder whether somewhere at the BMC — the Bulk Mail Center — it was written "*none* scheduled" instead of "non-scheduled," whether my father was mispro-nouncing it for his own amusement or mocking the men above him. "It's my nun-scheduled day," he'd say in the way he had of sounding facetious and formal at the same time. "We have the same nun-scheduled day," he said when introducing me to a woman, a co-worker of his, we met one day when we were shopping. "Oh, Tom," she gushed as if he'd said something profoundly personal. "Your father," she said, shaking her head. "Be ready," my father told her as we turned to leave, and she laughed. "Be ready for what?" I asked when we were sev-eral aisles away. "Be ready for the Lord," he said. "That's what the plates on her Cadillac say. RU-READY. Old Sister Prince. If they'd let her, she'd do baptisms at the BMC."

"Are you ready?" he kept asking as we shopped that day. "Are you ready, little girl?" he asked Kelly when we got home, and thinking he was referring to where they were going that night, she scowled, saying she had plenty of time. "Plenty a time," he told her. "Plenty. Too much time. But in the end you won't be ready. Five minutes to go, I'll be the only one dressed." And then he was off and running, as my mother de-scribed it, off and running, repeating some variation of those words all afternoon.

I waited for him to reappear on Sean's bike, the red flash of it carrying him around the corner. "Are you ready? Be ready." Was he that way at work? Did he talk nonstop as he sorted the mail, tell jokes, sing the way he did at home while he mopped the floors or vacuumed the rugs? "I could've been one of the band, one of the boys. A *sing*-er," he'd say if one of us walked past the room where he was working, and he'd take off the top hat we'd given him one year for his birthday, the one he always wore when he did housework or went to the bank, and bow. Was he like that at work? Did he amuse people or annoy them? His nun-scheduled day, I thought as I waited for him to come around the corner. What a difference it would have made if he had already started speaking that way when he picked me up at Northwestern. "Goddamn it, it's my only nun-scheduled day," he might have let out, pounding the steering wheel as he said it. Would I have laughed? Would it have eased the tension between us?

Down the street, he came into view and started riding in circles again. Sleek bike, I thought. Maybe that's why he wanted to ride it. Maybe it was the same urge that made Sean eager to fit into my father's speed skates. With their long, thin blades, they were the ice-skating equivalent of a racing bike, every aspect of them designed for the purpose of going faster. I remembered my father skating ahead of Michael, Mary, and me when we were children. Satisfied that we were okay on our own, he'd skate a few fast rounds, his hands clasped behind his back, his body bent forward. He looked almost artificial, like something that, though moving, could be stilled into a perfect form, a page from a picture book, a statue called "Man Skating." It was the way Sean looked riding a bike or running. I'd never thought of my father as athletic, though he had been a natural skater, with a flawless, easy style.

I watched him now, circling on Sean's bike. He was riding from one end of the street that ran into ours to the other. He was a man who liked repetition; maybe that's why skating had appealed to him, lap after lap. To me, the best part of going to the rink as a child was having my father tighten my skates, doing his best to make my ankles immovable. With his skates already on, he would stand before me, holding one of my feet and then the other between his knees, pulling the laces so tight there was nothing I could do but skate when he was finished. It wasn't just the snug feeling of my skates that I liked; it was watching him do Michael's and Mary's as well, each of us waiting on the bench until everyone was ready. Stepping out on the ice together, I knew that because of my father, all of our feet felt the same. Supported, he called it.

Why had he stopped skating? I wondered as I watched him circle the street. I should have gotten on a bike and ridden with him, led him out of the safety of our subdivision. He felt timid,

I could tell, about riding in traffic. But I had no desire to do it. Besides, it was flat down there. He was fine where he was. Eventually, he would have to ride back up the hill to our house or walk it. Tom, your heart. Your heart, Tom, my mother would tell him. "We're working twelve-hour shifts all month," he wrote in his letters. "I'm worried your father will have to go on disability. He's looking so tired. His heart . . ." my mother wrote in hers. But he would ride it out, work until it was time for him to retire.

I watched him make a big loop and head out again. It would be a few years still before that time came — retirement — a few years before I met Tiny Johnson and grabbed at the chance to find out what kind of man my father was at work.

"How's Tom took to bein' retired?" Tiny asked me. He was my seven-thirty student at the community college where I tutored on Monday nights. In the half-hour we had together, week after week, he hoped to produce a paragraph good enough for me to pass him into the next class on the long road out of remedial writing. When I learned that he worked days at the Bulk Mail Center, I asked if he knew my father. "Hell, sure," he said. "We worked off the same section. How's Tom took to bein' retired?" Each week, he would come in with a message. "Tell your old man we took care of Balinski. He'll know what I mean," he said one week, and he winked. "Really? What else'd he say?" my father asked, his eyes wide, when I relayed the message.

So was my father a thug at work? Or was Balinski — "some hotshot on the swing shift," my father called him — just the butt of a practical joke, my father one of the retired, perhaps revered, practical jokers? In the hope of finding out, I asked Tiny Johnson to write a paragraph about my father. "About Tom? Hell, that won't be hard," he said, and the next week he handed me a piece of paper with four sentences on it: "The thing I liked best about Tom was his thermos. It held six full cups of coffee. I asked him where he got it. Target he told me."

"Can I keep this?" I asked.

"Sure, yeah," he said, proud of his progress.

Before our session was over, I asked Tiny Johnson if he could tell me anything more about my father. "More than what's here?" he said, pointing to the paragraph.

"Yeah. I mean, I wondered what kind of man he was at work."

"He didn't die, did he?" Tiny asked, and then, as an afterthought: "He took the lump sum, didn't he, your dad? He took the lump sum's what I heard."

The lump sum or a distribution? My parents had debated it down to the dollar. Money — it had always been a subject of daily discussion at our house. Watching my father ride Sean's bike, I remembered the way he whistled when he heard how much it cost. "You're kiddin' me," he said. He lifted the bike with two fingers; it was that light. "More than my first car," he said, and he handed it over to Sean, who balanced it on his fingertips for a few minutes before pedaling away, my father smiling as he watched Sean disappear down the street.

Would Sean have become good enough to compete? To enter races, as he had hoped? How many jobs would he have had? How many things would he have tried before he became what he wanted? I thought of Michael. For a while, his "careers" were as many as mine. Cleaning office buildings, selling stereos, playing Bozo the Clown on a local TV show, making his way through the ranks at TWA — all precursors to his ultimately becoming a lawyer. I remembered how young he looked leaving for clown school, an opportunity that developed almost overnight when the TV station where he worked as a stagehand asked him to be Bozo. "They're sending me to the Larry Harmon School of Clowning in Florida. He was the original Bozo," he said when he woke me in the middle of the night to tell me he was leaving. He was kneeling beside my bed, whispering, listing all the things he would learn. I would also be leaving in a few days, returning to Northwestern for what would be my final term. I was nineteen; Michael was twenty-three. "I hope you make it this time," he told me. The moonlight fell across one side of his face, leaving the other in shadow, and it made me want to brush my hand across his forehead. He was tall and thin, and with his soft eyes and narrow nose, he looked like my father. "Why do you have to leave in the middle of the night?" I asked. "Redeye," he said. He kissed me on the mouth. "Don't screw up this time," he told me.

When I came home a few months later, Michael was working for TWA, his stint as Bozo having already ended. "The station lost their syndication rights," he said as he showed me the tricks he had learned, demonstrating the voice and laugh he had mastered. He pulled balloons out of his pocket and twisted them into animal shapes. "Make me a rabbit," Sean said. Michael tapped him — *abracadabra* — on the head with a baton-shaped balloon: "There. You're a rabbit." Sean was ten. "Very funny," he said, and Michael shrugged and smiled as he twisted the balloon into two long ears, a body, and a bunny tail. "It's the kind of skill that, once you learn it, you don't lose," he would tell me years later, when a fellow lawyer at his office asked him to play Santa at a children's party. "Being a clown or twisting balloons?" I asked. "Both," he said.

Like riding a bike, I thought as I watched my father. Was that the saying, or was it "like *falling off* a bike"? I would have to ask Michael. He was good with things like that. "I want to finish law school before I'm long in the tooth," he had once told me. It was the first time I'd ever heard anyone actually use the phrase "long in the tooth." "Shakespeare, isn't it?" he said when I pointed it out. With thinning hair, a full face, and a prominent belly, he had grown from a young, gangly Bozo to the perfect Santa for a children's party — beneath the fake white beard, a lawyer who represented people unjustly removed from their jobs.

Would the road to becoming someone have been similar for Sean? Would he have wandered around a while as Michael and I had, or would he have been more like Mary, who had moved directly from retail into accounting, or like Kelly, who always knew she would be a nurse? Who would he have become? How would he have achieved it? He had already been a paper boy. He had already bred tropical fish. Maybe he would have been a wanderer. Or maybe not; there's an inherent slowness to wandering, and he was always running, racing.

Cycling up the street now, up the hill to our house, my father was struggling. Shift, I said to myself, since he was too far away to hear me. *Shift,* I should have shouted. Instead, I walked down the driveway, planning to meet him halfway, and as I did, he got off the bike, waved to me, and began pushing it slowly up the street. His face was red when I reached him, and as we walked up the hill, I remembered what he told me the day he drove me home from college, when, after miles of silence, we finally reached our street. "Y'know," he said, and then he fell silent again. We pulled into the driveway and he turned off the engine. I started to open the car door. "Y'know, you feel bad now" — he touched my arm to stop me from getting out — "but life isn't always all or nothin'. Sometimes," he said, "it's a little here and a little there."

· 3 ·

After my father retired, he took a job working Tuesdays and Thursdays at the Earth City Auto Auction, where he helped wash hundreds of cars before driving them around an indoor track while the dealers — the "big money men," he called them — tried to outbid each other. He enjoyed the job, was good at it, and got promoted. Instead of driving the cars through the car wash, he recharged their batteries, receiving a raise of forty-five cents an hour.

He did it, he said, for the pocket money. For the first time in his life, he didn't hand his paycheck over to my mother. He cashed it and spent the money on whatever he wished, buying himself breakfast when he wanted to, taking drives in the afternoon — no destination — without worrying about how he'd refill the tank.

I imagined him driving across the river to Granite City, to the streets around the steel mill where he had grown up. ("We lived in the back room of that place once," he said, pointing out a storefront to me when I was a teenager. "It was a confectionery, and the man who owned it took us in for a while, whole family. Otherwise, I guess we would've been homeless. You'll hear your aunts tell it that my mother ran a candy store. Hell, maybe you'll even hear she owned it.")

He would stop at St. Elizabeth's, the church across the street from one of his many boyhood homes, sit in a pew and pray. My father's relationship to religion was a quiet, constant one. He carried a calm assurance into every church he entered, a kind of confidence, as if in each church, even if it was empty, he was welcome and well known. At some time or another, he had taken me to most of the churches that held any meaning for him. There was St. Ambrose, where he was married; All Saints, where Michael was baptized — his first son, his first sacrament of fatherhood; the Pink Sisters Chapel, where the cloistered nuns, wearing pristine pink habits, were separated from the rest of the congregation by an iron gate. It was there he went each June to make his novena. When I was a child, I thought it was this chapel, these nuns, his ritual of returning there each summer, that made him successful at selling brick.

He would drive through the industrial sites on each side of the river and then down Delmar, past the first apartment he and my mother lived in after they got married. He had put down deposits on several apartments, unable to decide which one would be best. Forfeiting the other deposits was the last such act my mother allowed him as an unbudgeted bachelor, she once told Mary and me, laughing. He would drive past the car dealership where he'd bought my mother her first car, a cream-colored, tail-finned Dodge. "Can't believe they're still in business. Got 'em down to nearly nothin'," he'd always say. If he were younger, he'd stop at a tavern, but years ago he'd traded beer for additional cups of coffee during the day, and years after that he demoted himself to decaf, so he was more likely now to choose an uncrowded place with a counter and free refills.

Occasionally, I imagined, he would stop by the cemetery. If I was right about the drives he took, he often passed Calvary on his way home. On holidays and birthdays, he and my mother went there ceremoniously, with wreaths, floral arrangements, and the familiar words they used when they talked to their son and to each other. Once when I went with them — it was the Easter after Sean was buried — we stood looking at the few blades of grass that had just begun to grow above his grave. The mound was still unmarked, and I asked my father why it was necessary to wait six months for the headstone. "The ground has to settle," my mother answered, filling in for my father with words that sounded — had he been able to speak them — like his own.

It was the ground, and what goes on in the ground, that got him out of bed the night before Sean was buried. He sat up until morning and called the cemetery as soon as it opened to ask whether it was too late to change his mind about the cement vault. The vault was optional but recommended and would act like a foundation, an insulation between the casket and the ground. It was not too late. He would like it then, he said, the best one, whatever the cost, and he spelled his name over the phone, telling the man Sean's age, how young he was, how he would be there so much longer than everyone else.

Other decisions had already been made. "Will the boy be buried by himself?" the funeral director asked.

None of us knew what he meant.

"You have the option," he explained, "of turning it into a double plot. It's customary for couples," he continued. "I know that here, of course, that's not the case. But some people have environmental concerns. I don't know what your feelings are about conservation. I'm sorry I have to ask. It's just that later, it's too late. We have to know ahead of time how deep to dig. Is there anyone in your family who might want to be buried above him? Someone who's close to him and single? An uncle? An aunt?"

"Everyone else will want to be with their husbands or wives," my mother finally said, crying. "I don't want you to bury him deep."

She didn't want to bury him deep, or in land that was unfamiliar. On the day after he died, she asked her family for one of the last remaining lots in the plot her great-grandfather had bought for forty-two people. That was her comfort, that the ground around Sean would be filled with family, like a solid bedding of inseparable bones; that kind, protective voices would be speaking to him through the soil: Here, Sean; Over here, Sean; I'm beside you; Take my hand.

That he would not be alone.

So there were those two concerns for him: my mother wanting him among relatives who could reach out to him, like the tips of tree roots touching and recognizing themselves as one and the same; my father wanting to shelter him from all the elements — the earth and air and water. Each wanting, really, for him to rise up. And theirs were not opposite intentions. They were my parents' individual manifestations of love: my mother and her love of nature, her flower gardens, the room she gives things to grow; my father and his brick and mortar, his faith in materials men make of the earth. And though it is in a way ironic, it did not seem ironic, then or now, that while my mother wanted Sean an arm's length away from her loved ones, my father wanted to keep anything from penetrating the place where his body was buried.

"The ground has to settle," my mother said that first time I went to the cemetery with her and my father. The ground settles. I held on to the thought as my father drove us home that day, none of us talking.

When he is not driving, my father spends a great deal of time painting and wallpapering. In one of the rounds of redecorating that began, leisurely, after the last of us left, and began again, rigorously, when my father retired, my parents started referring to each room by its color: the peach room, the yellow room, the mauve and blue room. Sean's room became the brown room, then went on to become the bear room, housing a collection of my mother's that grew so large it became a motif. Decorated with bear-printed wallpaper, pictures, pillows, and curtains, the room was filled with every type of bear imaginable — stuffed bears, porcelain bears, bears costumed for all kinds of occasions. At one point, I asked to move back to my parents' house for an extended time between apartments, and Sean's room was redecorated again to accommodate my stay. The bears were dispersed throughout the house — the great migration, Michael called it — and my father moved Sean's desk downstairs to make room for mine, a large library table with gargoyle legs that my mother said was definitely different.

I sometimes think it's Sean's desk that draws my father to the basement. It's the one part of Sean's life that remains much as he left it. Though a few things might be missing — I took his library card and a piece of paper on which he had calculated the capacity of all the aquariums he kept in the basement; he had written down the names of the species of fish in each tank, along with the average length of their bodies, posing a question to himself at the bottom: "Move neons to ten-gallon tanks, breed black tetras in twenty?" — the contents are largely intact. There's the Boy Scout badge he earned for swimming a mile; a map of Missouri; a travel-size chess set; a baseball signed by Lou Brock. There's an essay titled "Open Topic #8" that he wrote for English class. "When I think of all the good points of human nature," it reads, "and all of the negative examples, I would argue that the positive outweighs the negative handily." There's the card he carried in his wallet saying he is Catholic, call a priest. A church bulletin announcing that he'd won the eighth-grade cross-country invitational, setting a new record that would, for many years, remain unbroken. A blue ribbon from the science fair, four first-place ribbons for cross-country, a compass, a pocket knife, a key chain he made at camp, a collapsible Cub Scout cup, his coin collection, a field guide to the flowers of North America, and a pin that says *Peace*.

And there are things in his desk drawers that he never saw, like the letter that arrived a few days after he died congratulating him for making the honor roll; the hundreds of condolence cards that people sent; the poems girls wrote about him and read, crying, at his wake. Maybe they were even the girls he mentioned in the letter that he left, the ones who sat behind him and teased him, hoping he would turn around.

It was in the basement, sitting at Sean's desk, that my father finally rolled the change — all the quarters, dimes, and nickels — that he and Sean had saved for the trip they never took. He placed the rolls of coins in the bottom drawer, where they would remain, unspent, a little longer. The drawer was filled with coin wrappers, rubber bands, empty margarine tubs (Sean's system for sorting money), and the *Post-Dispatch* work apron he had worn each week when he stood at church and sold papers.

"I guess I should've given this stuff to Borgemeyer," my father said, naming the boy who had inherited Sean's newspaper business. My father had given Borgemeyer's name to Sean's boss, mainly because he had been one of Sean's best friends, but also — and this was in keeping with the way my father worked — because Borgemeyer had eight younger brothers and sisters, and my father figured they would be able to keep the job in the family for a long time, handing it down from one kid to the next. Aside from the apron, there was nothing in the drawer that any of the Borgemeyers would have wanted, and despite the decisions my father may someday have to make based on what he can fit into a footlocker, I imagine the apron is one item with which he will never part.

On the other side of the basement, opposite Sean's desk, my father has set up a card table that holds the pieces of whatever jigsaw puzzle he has in progress. Since his retirement, Michael, Mary, and Kelly have taken to giving him puzzles, increasingly complex, for birthdays and other occasions. My father has never bought a puzzle himself; just as, though he enjoys reading, he has never bought himself a book or borrowed one from the library. He reads only what is presented to him, books chosen by others based on what they have determined his tastes to be. When I first saw him poring over a puzzle, I wondered what prompted Michael, Mary, and Kelly to introduce him to this pastime, and then I realized that the idea my father shared with me — his idea of interlocking bricks — is one he must have shared with each of them, and that, watching him with the puzzle, I am witnessing its end result, the trickle-down theory of one man's dreams and ambitions.

The card table where my father does his puzzles is surrounded by furniture that all belongs to me. I am living in one room in New York, coming home once or twice a year to visit, and using my parents' house to store the possessions of a more spacious past. If my father were to add up what's mine in the garage, the bedroom, and the basement, he tells me, he figures he would find that I occupy close to a quarter of his house. What he does not say is that by accepting the accumulation of my life, he has added considerably to the encumbrance of his own. What he does not say, what he does not even suggest, is that he is sacrificing his future for my sake: while I live in one room, unencumbered, he lives in a house filled with boxes of my excess and indecision.

What he says instead, phoning one night, is that he has found something interesting in my desk drawer, and he describes what he calls a small wooden spyglass that, when he looks through it, makes multiple images of everything he turns it toward.

"It's some kind of kaleidoscope," he says.

"I have a few of them," I say, embarrassed to admit to him how many.

"Not like this one. All the others are in a box in the basement. This is the one you left in your desk drawer," he tells me, trying to lead me, long distance, to an exact location and letting me know, as I have always known, that he is a man who has nothing to hide, that unlike me, he does not feel the need to conceal his curiosities.

"I know which one," I tell him, imagining him looking through it as we speak. "It's faceted to simulate a dragonfly's eye."

"Is that right? Well, it's the wood on it that I like," he says. "I thought I might keep it in my room, if you don't mind."

"It's yours," I say, admiring his approach.

My father and I don't speak well on the phone, to anyone or to each other. Our conversations follow a cordial, unprovocative form, a checklist for keeping in touch. Yet we always hesitate to hang up, as if doing so will do much more than disconnect us.

"There's one more thing," he says.

"What's that?" And all at once I am anxious about what I may have left behind, concerned that he's uncovered something he considers compromising or questionable, that he is about to confront me with the evidence of something better left unsaid.

"Your brother's birth certificate."

I am taken aback by this. I remember immediately where I left Sean's birth certificate, and it reveals to me the depths of my father's discovery, all the photographs and letters that are stored in the drawer where I put it, all innocence and no innocence, like a path through my past. I could see in this a perfect reciprocity, the inverse relation of our lives, all the days I spent searching through my father's drawers, all the items I discovered that, given my method of discovery, denied me the opportunity of inquiring directly about them, about what part of my father's life they represented or why they were significant enough for him to save in a shoebox or under a stack of T-shirts in his second dresser drawer. It was a process that required patience, often unrewarded, the time that passed between finding some seemingly mysterious item and hearing the story — days, months, maybe a lifetime later — that provided it with a place and a purpose, an explanation that more often than not turned the most exalted discovery into something disappointingly mundane.

But when I think of my father, surrounded by whatever evidence I left him of my life, I suspect that his intentions were not the same, that he did not start out as I did — a child whose curiosity went unchecked — but rather went searching for some practical item that he had seen me use, a particular ruler or pen, some piece of printer's apparatus. Or perhaps he merely needed a piece of paper, and what began as a simple, utilitarian action became an afternoon of looking through the possessions of someone he thought had left him, someone he may have missed. And it is this — our desire to hold on to a part of each other — that connects us, turning our curiosity, whatever its source, into acts that are not idle.

And when, unlike me, my father wastes no time with his inquiries, I realize too that waiting is not just a matter of patience; it is a matter of perspective. I am young, and he is not. And for the first time I think of his mortality, of how we have been willing to wait a lifetime to reveal ourselves to each other, and I realize that my father's death, when it comes, however hard or easy, will be every bit as sudden as Sean's.

And at the same time I am struck by the fact that the words "*your brother's* birth certificate" represent a departure from the way my father usually speaks of the dead. Intentionally or not, he marked people's death by changing the way he talked about them. For people still alive he used *your*: "your mother, your brother, your grandmother, your sister, your uncle, your aunt," he would say to me when he was speaking about someone or telling a story. But when people died, they crossed over, in his mind, from *your* to *my*, so when he talked of his Aunt Annie she was no longer "your Great-aunt Annie" but "my Aunt Annie"; his brother went from being "your Uncle Bud" to "my brother Bud," and his parents, with whom I had no relationship, were never anything other than "my mother" and "my father." They always belonged only to him.

But he said "your brother's birth certificate," as if, in his scheme of things, Sean were still living. If I pushed him, maybe he would even say his name. "*Whose* birth certificate?" I could ask. "Your brother's. *Sean's*," he would have to say. The way he pronounced Sean's name had always amused us. We used to tease him and try to correct the way he said it.

"*Shawn*," we'd say, "like *Dawn*. Not *Shon*, like *Don*."

"*Shon*," he'd say.

"*Shawn*, like *lawn*," we'd say.

"*Shon*," he'd try again.

"*Shawn*," we'd say slowly.

"Ah, to hell with it," he'd say. "You people are all so smart. Your brother *Shon* doesn't give a damn what I call him. We have an understanding."

Now I'd give anything to hear him say Sean's name again, whatever way he wanted.

"So where'd you get it?" my father asks, bringing me back to the birth certificate.

"It's not the real one," I say.

"I know it's not the real one. The real one's filed in a box on the top shelf of your mother's closet; I checked. But it's not just a copy, either. It's notarized."

I remembered the feel of the notary seal, the circle of embossed bumps, as if I could read the birth certificate blind. The paper was black with white print to show it was not the original issue. I was shocked by this when I saw it, unprepared for the severity of white on black. It had a finality that at first glance made me think I'd been sent his death certificate.

Why had I wanted it? That was the real question my father was asking. Not where did I get it, but why did I want it?

"I ordered it."

"You ordered it?"

"Yeah. From the Bureau of Vital Statistics."

"Where are they located?"

"Downtown. I was there anyway," I say, "getting a copy of my own."

"Your own what?"

"My own birth certificate. I needed it for work."

"You go to great lengths, don't you," he says. I had heard him use this expression before. "That guy goes to great lengths," he had said about a man at the steel mill who was notorious for the lies he told. I was young when I heard it, and it left me with the impression that lying was measurable, like mileage. So I was unsure, in this instance, whether my father was commenting on the time it took to drive downtown or whether, in fact, he knew I wasn't telling the truth.

"It didn't take that long," I say.

I am embarrassed by what my father has discovered, and I have no explanation, for either of us, about why I ended up one day at the Bureau of Vital Statistics with absolutely no other business than buying a copy of Sean's birth certificate. I woke up one morning and went there. It was years after he died. Weeks later, when the certificate arrived in the mail, I read the facts. They were accurate, of course: the city, the county, the state, the size of his body at birth — all the indices of his existence — my father's occupation, my mother's maiden name. I can't remember now how easy or hard it was to get it, whether I just walked in, stood in line, and signed something, then went home and waited, or whether I had to furnish proof of my identity and offer the clerk a reason for why I wanted — why I needed — this retrievable record, this proof that he had been born.

I don't know why I wanted it. I don't know why I read it every day for weeks, or why, each time I picked it up, I ran my fingers over the notary seal, its texture rising above the surface like an affirmation of the truth. I don't know why I had to have it. I didn't expect to discover anything from it. So why? What could I tell my father? And why did it cause me such embarrassment? Why did I want to pass it off as an afterthought, wrap it around some practical purpose? I would have preferred that my father had found something else, anything else. I would have preferred not to have to confess to something I didn't understand.

When I was a child, preparing for the sacrament of penance, I practiced my first confession with my father. He played the part of the priest; I played myself, a seven-year-old who had been scrupulous about keeping track of the sins she had committed. I started a list when I was six and added to it all year in anticipation of this dark encounter, the day when I would have to enter the confessional and say, "Bless me, Father, for I have sinned."

I thought the practice should be as authentic as possible, so my father and I sat on opposite sides of my bedroom door. I left the door slightly ajar so we could talk through it. It wasn't dark enough, but it would do.

"Bless me, Father, for I have sinned," I said.

"How long has it been since your last confession?" he asked.

I poked my head around the door. My father was reading the newspaper as we practiced. "Dad," I said, "it's my *first* confession."

"Oh, right," he said. "Welcome to your first confession."

"Dad," I said, "that's not what they say."

"How do you know what they say?" he said, turning to the sports page. "You've never done it."

"They teach us in school what they say. Now start over."

"Okay," he said, putting down his paper, and he walked us back through the opening lines and then turned it over to me.

"I am heartily sorry," I said, following the prescribed form, and I took the list from my pocket and began spouting off my sins.

"Are you readin' this?" my father asked. He poked his head around the door. "Let me see that," he said, taking the list. "This is not a sin. This is not a sin either," he said, and he took a pen from his pocket and scratched one sin after another off the list. "What's this mean?" he asked about an entry near the end, and when I told him, he said, "That's not a sin for a seven-year-old."

When he handed back the list, only a few white lies were left.

"What do I do now?" I asked, worried that I wouldn't have enough to say.

"Sins of omission," he said.

"What?"

"They're all the things you should've done but didn't. They're called sins of omission. Just use it as a category," he said. "It makes for a quick, clean confession."

"Are you sure?"

"Sure," he said, tapping my head with his paper as he walked out of the room.

The next week, when I made my first confession, I told Father Toomey I had committed several sins of omission, and he asked me to elaborate, so I asked to start over. I pulled out my list, which I'd brought along for safety's sake, offered up the few white lies that were left, then tried to read some of the sins my father had scratched out.

"I watched my great-aunt get undressed," I told Father Toomey. "She thought I was sleeping." This was the entry my father had told me was not a sin for a seven-year-old.

"What was it you thought you would see?" Father Toomey asked through the screen-covered portal. It was higher than my head, but it was even with Father Toomey's and framed the left side of his face perfectly. He loomed, like a phantom profile, above me. When I finished my confession, he would tell me my penance, close the opening in the oak door between us, shift his weight a little, open the portal to the other confessional, and show his right side to someone else. It was my first time in a confessional, and though Michael and Mary had explained the mechanics to me — how the priest sits in the middle booth and extracts confessions from either side; how you don't start talking until the hole on your side slides open, leaving only a dark screen between you and the priest (Michael had issued all of his first confession before the priest had ever appeared) — I was unprepared for how indifferently the priest alternated between sinners, opening one side and closing the other, like a train conductor switching tracks. And in the darkness, instead of concentrating on my sins, I started to think of the confessional as a visual demonstration of a phrase my father was fond of: "in one ear and out the other." Then I wondered if thinking such a thing would be considered sacrilegious, and concluding that it was, I decided that I would save it so I could add it later to another list.

Father Toomey reminded me that he was waiting for an answer. "What did you think you would see?"

"I don't know," I said, and he started to explain the merits of making a full confession, but before he finished, he was interrupted by someone knocking on the screen next door.

"Stay put," he told me, and he turned his attention to the other confessor. A voice I recognized as that of the boy who sat behind me in school said, "Am I supposed to let you know I'm here?" With both portals open, I could see Father Toomey more clearly. He opened the curtain of the confessional, letting in more light, and I watched him take a quick count of my classmates who were waiting in line to confess. He was old, and the length of the line seemed to make him nervous.

"Say ten Hail Mary's before tomorrow," he said, turning back to me, and he shut the screen door on my first confession.

When I got home, my father asked me how it went.

"Okay," I said, and he handed me a small, palm-sized book called *A Summary of the Spiritual Life Simplified and Arranged for Daily Reflection,* which he said had been given to him when he was a boy. The chapters had titles like "Conquering Bad Habits," "Virtues Leading Directly to God," and "Self-Conquest through Mortification." I turned to the chapter on bad habits and saw that it contained a section called "Needless Curiosity" and another called "Curiosity of the Eyes," which seemed relevant to the confession I had not fully given.

The book was narrated by God, and each chapter opened with a direct address to the reader. "My child," began the section called "Curiosity of the Eyes," "many a sinful thought and desire was brought into your soul by your curiosity to see things. The memory stores up pictures of what is seen and in a sense makes these pictures a part of you. By a reasonable control of your eyes, external temptations will not reach your imagination so easily. I do not begrudge you a reasonable need to freshen your spirit with a nice change of scenery, but I warn you against too free and easy a use of your eyes."

I turned to "Needless Curiosity." "My child," it said, "uncontrolled curiosity can waste a good deal of time and energy.

It leads to pointless visiting and useless conversation. Many sins of omission and carelessness spring from uncontrolled curiosity."

Despite the reference to sins of omission, I could tell by the freshness of the pages that my father had never read the book.

"So why'd you want his birth certificate?" my father finally asks outright.

The silence, long distance, starts to add up.

"It's . . . I just . . . I don't know," I answer. I want to say to my father, Bless me, Father, for I have sinned. I want to say, Here, please, look at my list. I want him to tell me what category to put this in. I want him to read me the lines, to help me, to forgive me. I have sins of omission, I want to say. I have sins of omission with consequences too severe to confess.

"You didn't get the death certificate too, did you?" he asks. I hear the worry and hesitation in his voice.

"No."

"Or . . . ?"

"No," I say, knowing that in his mind he's running through all the records that exist for Sean, and that now he is referring to the autopsy report.

"I've never even looked at either one," I say. Given my history, he sounds surprised and relieved to hear this.

"I think they're insulting," he says.

"You think what's insulting?"

"Those pieces of paper. Not the birth certificate. I can sort of see why you'd want it. But the others. They insult me. What are they for? Proof? That he died? That they examined him? They just sent them to us in the mail, unrequested. Whenever I think about it, I get upset all over again that they had to do an autopsy in the first place, when he died right there in front of the doctors. Do you know how they autopsy a body? Do you know what they do to it?"

"Yes," I say.

"Why would they do that to a boy? To a perfect body?"

"I don't know, Dad. I never heard of it until then. The law about it, I mean."

"Unnatural deaths," he says. "Y'know, some of your relatives, I won't say which ones, asked to see the autopsy report. They wanted to know if Sean had been usin' drugs. They didn't think we were tellin' the whole story."

"I know. I heard," I say. "Try not to think about it anymore."

"Well, I think about him," he says. "I think about him until I just can't think about him any longer. Your mother always wonders what he'd look like. I tell her he'd be handsome. He was a good-lookin' kid. Everyone said. He was a good boy, a gentle boy. I liked that in him. I don't think he would've been really tall, though, like Michael. I think he maybe would've ended up just shy of six feet, like me."

"I'm afraid I'll forget his voice," I say.

"I can't remember his voice anymore. I lost his voice long ago."

"I'm losing it too," I tell him. "It's down to a couple of words — a sentence — I say to myself sometimes."

"What's the sentence?" he asks.

And as with the birth certificate, I find I am embarrassed to reveal this to my father, not the sentence, but the fact that I so often catch myself in a daydream, thinking of Sean calling it out, the way he made the words sound short and strong, saying them as if they were a conclusion rather than an announcement. I liked the way he said it, with that new lowness in his voice. The way he said it made me feel like a different person. With him, I was a different person.

"What's the sentence?" my father asks again.

"'I'm home,'" I say.

"'I'm home'? Hmm. Yeah." I can almost hear my father smiling and shaking his head. "Yeah. He always opened the door in a hurry. I could always tell which one of you kids was comin' home by the way you opened the door."

"Really? How did I open it?"

"Oh, the slowest. You'd open the door, and there'd be a long pause. You were always draggin' somethin' in behind you."

"Dad," I say.

"It doesn't matter," he says. "I just wondered why you had it."

"Not about the birth certificate."

"What, then?"

"Sins of omission."

"There are no sins of omission," he says.

"The night Sean died, I . . ."

"No," he says, "you don't have to . . ."

"It's just that I was supposed to call him," I say. "I talked to him the day before. I called to talk to Mom and he answered the phone, and he asked me if I'd call him the next night. I said sure. That's the last I ever spoke to him. I never called him. I was at work. I wasn't even busy, but I kept looking at the clock and thinking I'd call him later. I remember it being eight o'clock. I remember it being nine o'clock. I wasn't that busy. I don't know why I didn't call. All I had to do was call, you know; it would have been so simple. 'Sean, what's up?' And maybe he would have said something."

"And you think if you called he might still be alive?"

"No. I think if I called he would definitely still be alive."

"I think if I didn't have a heart problem," my father says, "if my heart medicine hadn't been in the house for him to swallow, he'd still be alive. Your mother thinks if she'd gone to bed earlier that night, if she'd gone upstairs to say goodnight to him sooner and noticed there was somethin' wrong, that he wasn't well, he'd still be alive."

"But she never goes to bed early."

"See what I'm sayin'?" he says.

"Yes," I say.

And soon after, when we finally finish our call, I imagine my father, miles away, undressing alone in the darkness, in the room he calls 2E. The night is not over. The night is never over, and when he lies down on his bed, the light from the hallway casts his own shadow beside him, and he looks like a man who is almost asleep.

Acts of Faith and Other Matters

It has always been my mother's habit to dust the house in total darkness. After everyone has gone to bed, she dusts all the furniture room by room — desks, dressers, chests of drawers — setting off small signals, sounds that no longer disturb us in the dark: the half-note of mechanical music from Mary's jewelry box when we were little, the successive jingle of drawer handles as she skims over them, faint rustlings that mark her movement through the house, entering our subconscious as the ambient sound of sleep.

My mother ends each day this way, dusting in the dark, and in the morning, as soon as she wakes, she dusts again, in daylight. She dusts in the late morning and the early afternoon. She dusts after lunch. She dusts before dinner. She dusts in the evening between TV shows, then ends her day again, dusting in the dark. Dusting, she will tell you, is one of the ways she deals with her depression. "I feel better when I dust," she says. "It's that simple."

Dusting doesn't work for me. It had been a bad year, with many episodes of depression. There were many doctors and many drugs, some causing no reaction, some causing bad reactions, none, so far, producing the right reaction. When I came home for Christmas, my parents, instead of just looking happy to see me, looked relieved, and when I went up to my old room, a few hours after arriving, I found that my mother had left a medal on my desk, along with a note:

> I bought this medal of St. Dymphna for you at the Catholic Supply. She is the patron saint of the mentally ill. I know people who have prayed to her and all have said it really helped. In their estimation, she's a really "active saint." I read the story about St. Dymphna in a book at the Catholic Supply, but I don't remember it. Maybe you can take some time when you're home to go up there and read it.
>
> I have been praying to St. Dymphna for you. Wear the medal or keep it in your wallet. I hope it helps to keep you above depression. A prayer to her when you think of it wouldn't hurt. Just something short like "Dear St. Dymphna, remember me," will do. Love, Mom

I left the medal on my desk and lay down on my bed. It was nearly dusk, a few days before Christmas, and in the window of my old room a green plastic wreath with white lights and a red bow hung dark against the glass. It was synchronized with all the other window wreaths to turn on, by means of an automatic timer, at six. I pictured how this would look from outside, the instant when everything happened, the white lights on the wreath in each window, the white lights around the lamppost and the porch. And I thought about how it delighted my father to have set up a system that depended on no one, that worked even in our absence, so different from the days when we were children, when one of us would go from room to room, lighting whatever there was to be lit, leaving a path that was obvious on the outside, one window and then the next.

When I woke up a few hours later, the lights of the wreath were warming the window. The windows upstairs were thin and wore the weather. Opaque with frost, they would shed their frozen skin in streams of sweat as soon as the temperature began rising. But for now, wherever the white lights touched the windowpane, tiny spots of frost had disappeared, making it look as if each little bulb had a halo.

I fell asleep again, and when I woke, I saw my mother dusting my dresser in the dark.

"What time is it?" I asked.

"About eight" was her answer.

She sat on the edge of my bed and wrapped her dust rag around one hand and then the other.

"Did you get some sleep?" she asked, and almost as soon as she sat down, she was up again, dusting my desk.

She picked up the medal to dust beneath it. "Have you heard of St. Dymphna?" she asked, but before I could answer, she apologized for having forgotten the story. "You know my memory," she said, and then she sat back down on my bed and asked again about the newest doctor, making me promise I'd stick with him until something worked.

We could hear the Christmas music that the man across the street piped out to his porch. The music had been a holiday staple for years. The first year, my father sent Sean over to the man's house to ask if he'd add "White Christmas," my father's favorite Christmas song, and when he did, we stood outside and listened.

"Pretty neat, huh?" my father said when it was over, and he smiled.

"Pretty neat," Sean said.

My mother and I talked for a while in the dark, the white bulbs of the wreath dimly lighting the room, the sound of Christmas music filtering faintly through the window. She was worried about me, she said, about how quiet I'd become, and just when I thought she'd said everything she had to say, she added, "You've always overwhelmed me. I've been in awe of you since you were young. I told God early on that I didn't think I'd ever have enough to give you. I told him you were the one I'd have to let go."

Her words clumped in my throat and I closed my eyes, feeling instantly altered somehow by this long-held assessment of me that she had never shared.

"I can't say it better, but you know what I mean." She leaned over and kissed me. "Come downstairs and watch TV," she said. "There's a special."

My mother loves specials, singing and dancing and variety shows. That night it was a Kenny Rogers Christmas.

"It doesn't matter as long as he's got good guests," my mother was saying in response to something my father had said before I came into the room. She was stretched out on the sofa, and when I sat down next to my father on the loveseat, he patted my leg.

"Kenny G," he told me, and he laughed as if he were about to say something more, but my mother broke in.

"Kenny Rogers," she corrected, telling him to stop talking, the show was about to start.

"Wreath go on in your window?" he whispered. "Pretty neat, huh?"

"Pretty neat," I said.

When the show was over, my father went to bed and my mother motioned for me to sit on the sofa. In our family, she's the only one who can stretch out on a sofa and not fill its length. I picked up one of her feet and put my fingers through her toes, spreading them like little nubs between my knuckles.

"Do people tan between their toes?" I asked her when I was a child. I was seven or eight and had just come home from school. Until I was twelve, my mother worked on her tan every day from May through September, lying out in the yard for a few hours in the late afternoon. I used to sit on the end of the chaise longue, still in my school uniform, and take out the papers in my bookbag, showing her each one before going inside to change clothes. Every day, she had something in a jar to show me — Michael and Mary too, if they were around, though usually, in a hurry to play with their friends, they just yelled, "I'm home," from inside the house and she would turn her head, squint at the screen door, and wave. Most days, it was just me and my mother.

In the month between May and the start of summer, she got darker every day, while Michael, Mary, and I remained as white as winter. By the time the pool opened, she would be fully tanned, and as we struggled and splashed around her, it must have looked as if she had come from somewhere far away to watch us, a stranger sent from the sun to keep our weak white limbs afloat.

"Do people tan between their toes?" I had asked, spreading her toes to see. I didn't really care, but it was the only part of her I could touch while she was tanning, the only part that wasn't covered with baby oil or cocoa butter or whatever she was using that year to coax the sun into her skin.

"I don't know. Do they?"

"It's hard to tell," I said as I looked between each toe. "I'll have to check again tomorrow."

Maybe there were butterflies in the jar that day. Maybe ladybugs or a larva attached to a leaf. Maybe a frog or a garter snake, a praying mantis, some crickets. Maybe there were mushrooms growing somewhere in the yard or a turtle she'd put in a box to keep it from wandering off. ("Buy a box turtle?" she had said when Allison told her how she planned to spend her First Communion money. There was a long waiting list for box turtles at the pet store — they sold for thirty dollars each — and Allison was ascending it slowly, hoping to buy both a male and a female so she could breed them. She would sell the babies, she said, for half the pet store's price. "Buy a box turtle?" my mother had said, expressing her disbelief, her sadness, really, that backyard wildlife could have disappeared so completely between just one generation and the next, her daily specimens, the things she delighted in showing us, becoming endangered species in the suburbs of St. Louis.) Maybe there were birds' eggs from an abandoned nest, or a spider so huge and frightening I wouldn't touch the jar for fear it would break through the glass. Maybe there was a centipede. Maybe there was a wasp or a bee or a yellow jacket my mother had caught so we could look at it up close without getting stung. And there were caterpillars, of course. There were always caterpillars. They were her favorite form of insect.

Lying on the sofa now, my mother pulled her toes out from

between my fingers. "Don't," she said. Her feet were short and callused. They had been longer and thinner once, and soft. Over the years, her legs had grown shorter too. Bare beside me, they stuck out from her knee-length nightgown. I laid my hand on her calf, and she let it stay there. Her leg was covered with freckles and beige-colored blotches. She had the skin now of someone who had tanned too much. I had the same kind of skin on my shoulders, but it wasn't from tanning. At the start of every summer since I was six, I'd let my back burn until it blistered. The first summer, the blisters surprised me. They popped up like tiny bubbles all over my shoulders, then grew together into yellow, oozing sacks before turning into scabs. "Don't do this again," my mother said, covering me with lotion, but until my early teens, I started every summer the same way, staying out in the sun unprotected, a temptation I continually succumbed to, like a special form of sin.

I was thinking about what my mother had told me earlier that night. I had overwhelmed her. Always. I had *always* overwhelmed her. She had to let me go. What had she meant? All my life, she was the one who defined me, who named my qualities — the good qualities and the bad — as if I would never see them myself or realize, without her words, who I was. I was hard on my clothes, tall for my age, quiet and clever. And I was different. That was the last thing I let her tell me about myself.

I was seventeen when she said it. She had spoken the words lightly, laughing, not knowing how heavily they would land upon me, not hearing what I heard. The words would stop me, silence me, still me. They would separate me from myself. It was a Saturday morning. She was standing at the kitchen window when I went to her, held her hands behind her back, and began kissing her and tickling her the way Sean and I often did, usually in the evening after dinner, making her squirm and scream with laughter until we let her go. It was sometimes this, the tickling and kissing; it was sometimes the disco dances that we dreamt up and made her do.

But that day Sean wasn't with me. It was morning. I don't

know why I did it. It wasn't usually the kind of energy I had alone, but I was almost overcome with happiness. I had been accepted to Northwestern, with a full scholarship, and I was nearly giddy with the thought of it. But my mother had become depressed when I received the news. After she read the letter, she didn't look at me or talk to me for several days. Each time I tried to approach her, she held up her hand and shook her head. Within a few weeks, though, she accepted the fact that I would be leaving, and she embraced my happiness, enjoying my excitement as if it were her own.

It was during those days, those heightened, exuberant days, that it happened. I was kissing and tickling her at the kitchen window. She was laughing and squirming and turning red, just as if Sean had been there with me. But this time, between bouts of laughter, she said, "It's one thing your doing this with Sean, but you're seventeen. I hate to think if someone looked in this window and saw you right now. They'd think there was something wrong with you." She laughed and tried to tickle me back. "They'd think you were — let me go now, I have to get to the bathroom — they'd think you were, you know, different." When I let her go, her face was the same red it always was when she laughed; she sighed in the same way she always did after Sean and I stopped; and as always she hurried into the bathroom, still laughing. She acted as she always did, but I was different.

Being different, I inferred from my mother's words, was something that should be kept out of sight, away from the window. Being different was something you wouldn't want anyone to see. I was seventeen, and I would hold off being really different until the day after Sean died, with Ellis, loving her then and all the summer after.

"What are you thinking about?" my mother asked, poking me with her foot.

"Nothing, really."

"How do you like the Christmas tree?"

"I hate it," I said, and she laughed.

With all of us gone, my parents no longer put up a real, ceiling-high Christmas tree. Instead, they had a small artificial one on a table next to the TV. Beneath it was the manger scene they bought when they got married, the pieces placed in odd positions by my nieces and nephews. Today, all the animals were inside the manger, all of the people were out, and the angel was face down on the floor, several feet away. Tomorrow, Christmas Eve, there would be another arrangement, and still another the next night, each visit by my nieces and nephews yielding a new interpretation until the day it all went back into a box in the basement.

The figurines made me remember all the Infant of Prague statues my mother had given me over the years. I was running out of places to put them. A statue of prosperity, it had to stand facing the door. "Why does it have to face the door?" I asked when I was a child. I was helping my mother dust, and I'd turned one of the statues toward the other side of the room. "So if success comes knocking at your door, someone will be there to see it," she said, and she turned the statue to face the front of the house, where our good fortune was sure to arrive any minute. My mother slipped an Infant of Prague into my suitcase whenever I went anywhere, sending me new ones periodically to remind me of the power of devotion to this Czechoslovakian incarnation of Christ, a child ("Why do they call it an infant? It looks more like a toddler to me," Sean once said) robed in riches, wearing the crown and cape of a king and holding a globe in one hand and a cross in the other.

When I moved to New York, I sublet a studio sight unseen from a woman I met one summer. A thick layer of dust covered the walls, the floor, and every cluttered surface. Just to find a place to set down my luggage, I had to dig through piles of debris: papers, water-stained photographs, dirty clothes, a moldy swimsuit, several broken umbrellas, all of them black and harboring bugs — things that I could barely bring myself to touch. The most alarming was a soiled paper plate on which the words "May 5, 1968, 11:08am, Mom dies" were written between crusty streaks of dark residue. More than anything else, the paper plate made me stop and question what I'd done, moving so far away from my family to live in a room I'd be ashamed to let them see.

When I had cleared enough space to unpack my luggage, I found that my mother had put a statue of the Infant of Prague in my suitcase. Trimmed in gold leaf, it reminded me of the porcelain statue of the Virgin Mary she had wrapped up and given to me for Valentine's Day one year. It was my father's wedding gift to her, and she thought I should have it. Each time she slipped me one of her sacred treasures, she did it secretly. "Don't tell your sisters," she'd say. Each time, I let them find out.

That same day, the day I moved to New York, I discovered a pizza parlor around the corner with a larger-than-life statue of the Infant of Prague in the window. It was late at night, and I literally gasped at the sight of it. I had seen plenty of Infants of Prague in my lifetime, but this one, at least six feet tall, was lit up and looked like a beacon in the window. My first night in New York, I took it to be a message from my mother.

Now, six years later, I no longer lived in that studio, but I visited the neighborhood often, and the last time I passed the pizza parlor, I saw that it had gone out of business. Not even the giant Infant, turned properly toward the front door, had been able to insure its success.

"I have some bad news for you," I told my mother.

"What bad news?" She was swinging her legs off the sofa in an effort to sit up, pulling on her nightgown so it wouldn't rise up too high on her thighs. "What bad news?" she asked again when she was seated. I put my hand on the part of the sofa where her legs had been. It felt warm and familiar.

"The Infant of Prague Pizza Parlor went out of business."

"Go on and make fun," she said. "Prosperity can mean many things, not just money."

"Yeah, but that was a really big statue, and the place went completely under."

"Maybe they didn't make good pizza. Did you ever try it?" she asked, and she got up and went to the bathroom.

There was an Infant of Prague in the bathroom, in the

kitchen, the hallway, and every room of the house. Some had clothes that came off, silk capes and crowns. Some were made of wood, some porcelain, some plastic. Though they were numerous, they were not conspicuous. They were well placed, pulling prosperity into our house, something that meant more than money.

As I waited for my mother to come back, I looked at the tiny Christmas tree, the figures of the nativity set arranged freestyle around it, and thought about the Infant of Prague Pizza Parlor. Beyond its size, the giant statue in the window had a disturbing effect on me that I couldn't explain. And then one day, a few months after I moved to New York, I was walking past it and the memory of another statue surfaced, an Infant of Prague I had seen in my grandparents' garage.

The garage — a dark, cool vault of stone and cement that ran beneath my grandparents' apartment building — was a favorite place for Michael, Mary, and me to play. The big clean cars of old people were lined up like luxury liners from another era, beautiful and gleaming, most of them set up on blocks. We played Hide-and-Seek between them, and under them, and in them. Other games took us into the many chambers of the basement, rooms that were damp and dark. On one side, padlocked storage lockers lined the wall. In the darkness, we could see things poking through the picket-style fences that enclosed them. One day we saw a head lying on the floor of one of the lockers. Through the slats of the fence, we could make out a nose, a chin, and a mouth. Michael took a flashlight from one of the cars, and when he shined it on a life-size statue of the Infant of Prague, I screamed. It was there, in some part of that subterranean space, that my mother's father shot himself when he was sixty.

My mother came back from the bathroom. She would be ready to talk now for a long time, and she began by telling me about the IMAX movies she and my father had been going to see at the new science center in St. Louis.

"What's good about IMAX movies," she said, "is that most of them have no story. There's nothing, really, that you need to remember from one minute to the next. Mostly, you're just surrounded by life, thrown into the middle of some kind of . . . what do you call it? The desert, the ocean — some kind of environment. That's not the word . . ."

"Ecosystem?"

"That's it. And it just exists all around you, and you get to witness everything that happens in it. It's spectacular, that's all I can say," she said, and then she proceeded to describe all the IMAX movies she and my father had seen, the ecosystems, the life and lands outside St. Louis.

I'd never seen an IMAX movie, but I imagined that part of its appeal was the relative lack of language, nature without an added narrative. As I listened to my mother, I remembered the bliss of being in the back yard with her when I was little, how much there was to see that I would never have noticed without her, how much we experienced together, how little we talked. As she tried to recreate the magnificence of an IMAX movie for me, I realized how essentially different her experience of the world was from mine. She had no urge, no inclination, to find the connective tissue, to order the elements into a coherent shape, some sequence that became a story, and I wondered if this almost total indifference to whether things had a beginning, a middle, and an end was one of the reasons her faith flourished. "Why?" was not a road of inquiry that interested her. She lived in the wide, wide world of God is good, God is great, a world in which all existence was equal and enthralling. She saw things grandly and minutely at the same time, a kind of vision that seemed similar to the IMAX scale of large and small.

She was talking about Antarctica, the subject of the latest movie she had seen.

"I've always wanted to go there," she said when she finished.

"Are you serious?" I had always wanted to go to Antarctica too, but it was the last place that I could imagine my mother going. She fell a lot and was afraid of ice. "Antarctica?"

"Yes," she said.

"It's probably pretty slippery."

"Oh, I don't know. I think a surface that's all ice would be less slippery," she said. "I always thought that ice here was slippery because it surprises you. But if ice is all you ever walked on, you'd probably get used to it. I think I'd fall a lot less down there than I do here."

A few days later, my mother would fall and it would be my fault. It was the weekend after Christmas, and we had gone to Mary's lake house in the Ozarks. When Mary suggested taking a walk on the nature trail at Ha Ha Tonka State Park, I waited until my mother left the room.

"Mom can't do that," I said.

"Why not?" Mary asked.

I had never been to Ha Ha Tonka State Park, but I knew from experience that nature trails and my mother were not a good mix.

"Well, for one thing, it's winter. There's snow on the ground. It's probably slick."

"Most of the snow has melted," Mary said. "If she didn't want to do it, she'd say so."

"No, she wouldn't," I said. "For some reason, unlike the rest of us, she has no idea what she can't do."

"What are you talking about?" Mary asked.

In the other room, my father was already chanting "Ha Ha Tonka" as he helped my mother put on her boots. My mother is overweight, has weak ankles and bad knees, and the upper part of her body bends forward considerably from her waist, making it look as if she could topple over at any time. Each Christmas, when I come home, I find her a little more stooped, and she slaps at me, laughing, when I sneak up behind her and pull back her shoulders to straighten her up.

"Lead with your stomach, Lois," I tell her.

"Quit. Quit," she says, swatting at me and turning red.

"Lead with your stomach, Lois. That's all I ask. Lead with your stomach, or you'll be walking parallel to the sidewalk soon," I say, keeping her from squirming away.

"I don't want to lead with my stomach," she says. "It's my worst feature."

"Who's going to do this for you when I'm gone?" I ask.

"No one, thank God," she says. "All the rest of my children know how to leave me alone."

"Someone has to straighten you up," I tell her.

"Oh, no, here we go again. Okay," she says, laughing, the next time I do it and the next. It's become our Christmas ritual.

At Ha Ha Tonka, we took turns walking with my mother. When it was my turn, I realized I'd been mistaken. The trail wasn't really a trail; it was a planked walkway built along a thin strip of woods that separated the lake from a steep rise of what looked to me like limestone. We had to stop a lot for my mother to catch her breath, but overall it was a manageable endeavor. Because my mother is short and stooped, she usually focuses on things that are low to the ground, and because she walks so slowly, not much escapes her notice. There was winter moss to see, its green turned to silvery gray. There were various forms of fungus. There were spiderwebs that had been spun on drier, sunnier days. Wet with winter now, they stretched between leaves and rocks and tree roots like glistening canopies, icy and abandoned. My mother wondered what it would feel like to live beneath those webs, to be a small animal or insect whose view of the sky was muted or dissected or otherwise altered by what looked to us like a beautiful gloss on the ground. I had never thought about another species' relationship to nature, about nature's relationship to nature. It would never have occurred to me to consider the sky from another creature's perspective, and I felt suddenly ashamed of my inability to imagine or appreciate anything outside my own experience. It was arrogance on my part. It was one of the many ways I was not like my mother.

Though we both wanted, someday, to see Antarctica. Why had I always wanted to go there? It was something about the starkness of the landscape that appealed to me, something about a vast emptiness, ice upon ice, whiteness upon white-

ness, a place devoid of soft surfaces. I imagined a wide, uninterrupted view of endless frozen space. I knew little about
Antarctica — I hadn't seen the IMAX movie — and I had only
the vaguest memories of grade school geography. It didn't matter. It was Antarctica in the abstract that I was after. It was my
constant state of self-exile, my need to get farther and farther
away from any kind of connection. It was the happiness I felt
as a child lying down in deep snow at the end of the back
yard. No one could see me. I was far from the house. It was
cold and crisp and quiet. Near dusk, the lights would start to
spill out from the houses and spread over the snow, making it
sparkle like a live white presence all around me, and I would
lie there, deep within it, the snow around me, the sky above
me, until the afternoon ended, five o'clock, six o'clock, the
dark end of a winter day, and I would hear my mother call us
in for dinner, call Michael, Mary, and me, first out the front
door and then out the back, her voice coming to me as if it
were the end of a long echo. I would lie there listening until she
called out again, until one of our names would be missing,
until there were two of us instead of three, my name again and
Michael's, or my name again and Mary's, until there was
only my name, the last one left, and I would lie there a little
longer, and when I knew there was nothing else, I would get up
and go in.

It was that long, empty echo, that clean, cold, quiet corri-
dor in which my sensations were both sharpened and dis-
torted, that place of peace and alienation, that I associated
with Antarctica.

What was Antarctica for my mother? Why did her desire
to go there surprise me? When I was a child, my mother's attitude toward winter was distant and remote. She preferred
to watch it from the window. My parents were seasonal in
some respects. In the summer, my mother shined and my father
faded. In the winter, it was my father who came to the fore,
taking us skating on Sundays, taking us sledding after supper,
building us snowmen, igloos, barricades for waging snowball

wars, the kids on our side of the block against the kids on the other. In winter, our back yard was like an amusement park of snow structures, my father the master builder, each of us his eager apprentice. In our lives, what my mother was to swimming, my father was to snow.

So it surprised me that my mother's long-desired destination was Antarctica, a continent of constant winter, but it also made me wonder whether people are drawn to the sources of their depression, whether there's a kind of dangerous comfort in them that we crave. My mother's depression was always worse in winter. So was mine.

"Maybe it's because her dad died in December," Mary said once when I asked her why people got sadder the closer it came to Christmas. I said *people,* but Mary knew what I meant. We were children then; Sean hadn't yet been born, and the worst of my mother's sadness was still around us. A few years later it lifted, and though it would never disappear entirely, over time it dwindled into much smaller doses of despair. Even now, with Christmas falling between the time of her father's death and Sean's, my mother manages, for the most part, to remain happy during the holidays.

Two days after Christmas, walking in Ha Ha Tonka State Park, her happiness was apparent. She was smiling and saying, "Good morning," to everyone who turned sideways to squeeze past us on the trail.

On our left, the lake was frozen at its fringes, and a few birds were skittering across its surface, pecking at a nearly bare branch of winter berries that hung over the bank. I knew my mother would want to stay there a while watching the birds, so I joined my father and Mary, who were walking ahead of us. "I'll be back," I said, and I hurried to catch up with them, hoping to hear more of the story my father was telling. I knew he would be reluctant to repeat it once he'd told it to Mary, and even if he did consent, I knew he would skim over it, giving me only the bare minimum, the bones. To make it worse, I could see that Mary was only half listening. Like my mother, she was more interested in the water and the woods and the great wall of limestone that was leaning closer and closer into the trail.

"Where's Mom?" Mary asked when I caught up with them.

"Watching the birds," I said, and I proceeded to ask my father about the parts of the story I had missed.

When I turned to go back, I saw my mother was on her hands and knees, with a small group of strangers gathered at her side. My heart sank. It wasn't guilt I felt but embarrassment.

Instead of running to her right away, I called to my father. "Dad," I said, "I think Mom's fallen." He ran back down the trail, Mary right behind him. I followed slowly, as he, Mary, and another man struggled to lift my mother to her feet.

"I'm okay," she said when I got there. She was standing up by then, and the people who had stopped to help her had already passed me on the path.

"I guess we should go back to the car," Mary said.

"No, I'm fine," my mother said. "Let's get to the place you wanted us to see."

Eventually, the walkway became a long zigzag of steps ascending the hillside, but just before that point, it widened into a wooden deck that overlooked a spring. It was the spring that Mary wanted us to see, and we continued toward it slowly. I walked slightly ahead, alone, my father behind me, and Mary walked with my mother. I could hear my mother pointing things out to Mary, the two of them stopping for a moment to discuss whatever it was — a rock, a leaf, some winter bud or blossom.

At the spring, the water was still and clear. Its surface mirrored everything above us: the cliffs, the clouds, and, in the center, the ruins of a mansion that had once stood on the highest peak. Only the massive chimneys and parts of the stone walls remained.

My mother leaned over the railing, and her face and upper body looked back at her from the spring. I was standing off to the side, and in the water I saw her waving to me.

"Come see yourself," she said.

"No, thanks."

"Oh, come on. It's fun to see your reflection."

"I already know what I look like," I said.

"No, you don't. Come see." And she pulled me toward her. My father and Mary had already seen themselves and were starting up the steps. They wanted to get a look from a landing that jutted out from the cliff about fifty feet above us.

"Will you please stay with her this time?" Mary said as she passed me.

"Look," my mother said after she had managed to make me move closer. In the water, her reflection looked bright white, her hair like a cloud that had been floating above us, stopping at just the right moment to frame her face.

"Don't look at me, look at yourself," she said, and when I turned my eyes toward my own reflection, it surprised me. It wasn't as bad as I thought it would be.

"See?" my mother said. And then she told me a story — I had heard it many times before — about a woman named Gloria Hall who lived two houses down from us when I was little. One day at a birthday party in the neighborhood, when all the other mothers were telling my mother how cute Mary was, Gloria Hall whispered into my mother's ear, "Mary *is* cute, but Kathleen is beautiful." Whenever my mother told me the story, she would whisper the words in my ear, as Gloria Hall had in hers, and then she would kiss me, her mouth still pressed to my ear, and whisper that it was true — I was beautiful. But now I was taller. When my mother stood up straight, which she rarely did, she only reached my shoulder. This time, it was her reflection telling me the story, and she went on, as she always did, to say how it pleased her that two of her children — Sean and I — had inherited her thick hair, and how on a sunny day the red and blond highlights in my hair shone like gold, and it was one of the most magnificent sights she had ever seen. She continued to list all my attributes — my eyes, my hands, how she marveled at my hands and Sean's, how she knew from their perfect shape that we were both destined to be not only kind but talented too. I listened, as I always did, still wanting, in my mid-thirties, to believe her, and as we gazed at each other in the water, a tear started down my cheek. Though she couldn't see it, I felt she knew it was there, knew why it was there, the deep shame I felt in betraying her by my embarrassment, frozen by my own self-consciousness when she fell. I tried hard to see myself as my mother saw me, our faces floating beside each other. I could see her beauty, but I could not see mine.

When I looked beyond our faces to the reflection of the rest of the world around us, I saw the bodies of strangers at the far edge of the spring, and when I looked up the side of the cliff, I saw my father and Mary among them, waving. As I waved back, I wondered whether my mother had ever told Mary a similar story about Gloria Hall, who agreed with all the other neighbors, saying, "Yes, Kathleen does get good grades, but it's Mary who's the smart one," and I knew that if my mother had told Mary that story, Mary must have believed it, because in many ways she had made it come true. When I stopped waving and looked back at the water — at the cliffs and clouds and ruins, at my face and my mother's — I appeared a little less ugly to myself, and I smiled at my mother's reflection, and the water showed her smiling back at me. And as we waited for Mary and my father to return, we talked about the ruins of the house that sat high up the cliff. We wondered whose house it had been and what had happened to it, why some parts had survived and others hadn't, why whole sections of walls were still standing while others, built at the same time, from the same stone, had disappeared completely.

As we made our way back to the car, my mother's breathing became labored, so we walked more slowly, and every few feet she would stop and let out two long sighs. *Hahhh, hahhh,* she would sigh — it was the natural sound of her sigh, the way she sighed wherever we went. *Hahhh, hahhh.* After a while we all started laughing, because the state park we were in was named Ha Ha Tonka, and my father and I began letting out a low bellow after my mother's sighs: *Tawwwn-ka. Hahhh, hahhh,* my mother would sigh. *Tawwwn-ka,* my father and I would follow, and people looked at us as they passed, and we laughed and walked on, wondering why the way back seemed so much shorter.

Hahhh hahhh, my mother sighed now as she rocked her body up from the couch. "Do you want some ice cream?" she asked, and she went to the kitchen and came back with two bowls of peppermint ice cream topped with chocolate syrup. Peppermint was my favorite, and my mother stocked the freezer with it whenever I came home for Christmas. We always ate it together in the middle of the night.

It was almost two in the morning. My mother had just finished describing the movie of Antarctica, and now she was telling me about a man, a cross-dresser, who came regularly to her garage sales. When I was growing up, garage sales were a common occurrence at our house. My mother held them often, converting what we would otherwise get rid of into a ready source of cash. Usually, the money went toward ordinary expenses, those that already occupied a place in her monthly budget, but sometimes she held a special garage sale, or a series of sales, to pay for something out of the ordinary, something for which we needed a lot more money — a superfund, she called it. The summer before I left for college, she held a garage sale every other week, and by September she had enough money for her, my father, Sean, and Kelly to spend a weekend sightseeing in Chicago after they'd delivered me to my dorm. It was Sean and Kelly's first vacation, and it made my mother feel good to combine the practical and the difficult — driving me to school and saying goodbye — with something frivolous and fun. Showing the kids Chicago, she said, distracted her from the loneliness she felt in leaving me there. Though garage sales no longer played an important part in her effort to make ends meet, she still found pleasure in having them; she could be counted on to hold several every year. The cross-dresser had only recently become a customer.

"He's a friendly man," my mother was saying. "And he

seems to really like my taste in clothes. Of course, since I'm big, a lot of my things fit him."

"But you're short."

"So is he. Isn't that surprising? I always thought of cross-dressers as being taller."

"Does he come dressed as a woman?"

"No. The first time he came, I thought he was buying clothes for his wife, and I found it touching, you know, that a man who maybe didn't have much money was willing to go to garage sales to find clothes for his wife. And then I thought maybe his wife was an invalid or something. You never know."

"How did you find out he was a cross-dresser?"

"After he came a few times, we started talking, and he told me he loved my garage sales because it was so hard for him to find clothes in his size. I felt kind of honored that he'd be that open with me, and now when he comes, he asks how I think certain things will look on him — you know, dresses and blouses — and I tell him, and sometimes he even goes into the back yard and slips them over his own clothes so I can say for sure one way or the other. Of course, he doesn't come back into the garage that way. I just look out the back door and tell him."

"Wow!" I said. "Is he a young guy or what?"

"He's middle-aged. Somewhere in his forties, I'd say. You know, before I met him, I never thought about cross-dressers. But it must be a difficult life to lead. Unless they're really bold, it must be so hard for them to get what they need."

"So your garage sales are providing a social service."

"You laugh, but it's true. It makes me feel good."

We were quiet for a while, and my mother asked what I was thinking.

"I was just thinking about that time in eighth grade when Sean and his friends dressed up in the girls' cheerleading outfits."

My mother smiled. "It surprised me that he did that; he was so self-conscious. It made me really happy to see him do it."

Sean and his friends had gone to cheer for the girls' basketball team, in appreciation for all the cheering the girls had done for them. It was the year Sean's team had gone all the way to the finals. A year later, when he died, the cheerleaders cried at the side of his coffin, standing together as if they were still a squad. They weren't wearing their outfits, of course, but I thought, as I watched from the other side of the room, that even out of uniform they looked like cheerleaders. No matter what they were doing, they seemed to do it in unison — crying, talking, touching their hair, maintaining their squad formation as they moved slowly away from Sean's coffin.

We were quiet again and I kept thinking about my mother's response to the cross-dresser. It surprised me and made me wonder if I had been wrong all these years about what she had meant when she said people might think I was different if they saw me tickling her. It was one thing to have done it with Sean, she'd said, the two of us tickling her together. It was a whole other thing without him.

It *was* a whole other thing without him. Life had been a whole other thing without him. It had been . . . well, it had been different. I remembered Ellis once saying she wondered who she might have become if her father had not died when she was six, whether she would have been someone different, someone other than an artist. I wondered the same about myself. I thought about who Sean would have become if he had lived, but I wondered too who *I* would have become. If he had not died when he did, I would be different. Who would you be? Ellis had asked. I would be me, I told her.

"Are you okay?" my mother asked.

"What?"

"Are you okay?"

I looked at my mother. What *had* she meant about my being different? I had never asked her about it, but I had let it come between us, creating a slight but significant separation. Had one of us done the same to Sean? Had we inadvertently said something to him that caused him to do what he did, something that altered the way he saw himself? One word — was that all it took? Uno, I thought. What if that was all it came down to?

"I'm fine," I said when she asked again.

She waited a moment, hoping for more.

"Really. I'm fine."

"Well, we better go up then," she said. "It's a busy day to-morrow — Christmas Eve."

I put the ice cream dishes in the sink, kissed her goodnight, and went upstairs. She would follow, but not right away. There was food to bring up from the freezer. There was cereal to set out for my father and furniture to dust. She would end her rounds in my room, dusting my desk and my dresser, dust-ing the chest of drawers, and when she was done, she would look down at my face as I lay in the dark, not to see if I was sad, not to see if I was the one she should watch, only to see whether I was asleep or awake, only to see if I was waiting for her, if I was ready, finally, for the real conversation between us to start.

· 2 ·

I was thinking about suicide. I had started thinking about it when I was fifteen, the age Sean was when he died. I was in ninth grade, and I began staying home from school a week or two at a time. I said I was sick, and I slept a lot. I was blessed with unnaturally large tonsils, so my mother believed me whenever I said I had a sore throat. Sometimes the doctor would check me for mono or strep, and when it looked as if things might start costing my parents money, I'd get better.

In my senior year, a boy brought a gun to school and quietly threatened to kill himself. He was suspended for bringing a gun into the building. I never saw him again. Like me, like Sean, he had always gotten good grades. If you got good grades, you could fall apart and no one would notice.

"It's the first inclination that's unnatural," a psychiatrist once told me. "It's that first time in your early adolescence when, for some reason, you feel you can't go to school the next day, and death presents itself as an option. If you've grown up in a family where depression is prevalent, it's possible — it might even be probable — that instead of being shocked or frightened by the thought, you accept it. You find comfort in it. It feels like something close to home. In effect, you give it room to grow. It becomes a daily message. You have this loud thought in your head that you've grown used to and accepted, and you live your life against it every day. It doesn't go away, or if it does, it's not for long. You'd probably be surprised to know," he added, "that some people go through life without having a single suicidal thought."

He was right. I found it hard to believe.

In my late teens and twenties, my depression only got worse. I dropped out of college. I went to work. Sean died, and I despised myself for it. The pain he felt, the pain none of us noticed, was always raw within me. There were weeks when I didn't sleep a single night. I showed up for work later and later each day. I kept stopping and starting over. I never got farther ahead. Because I didn't, I had no money. Because I had no money, I feared the future.

In more recent years, through my early thirties, except for a few intermittent good months, I felt hopeless. Each day, if I got to work by ten, I thought I could make it, but by evening I'd feel suicidal again. Each night, I went to sleep earlier, just to survive.

The psychiatrist I was seeing then said that people who inherit depression "come by it honestly." Somehow that made me feel less ashamed. "But you have your own history to be afraid of now," he said, "not just your family's."

In the fall of that year, the year my mother gave me the medal of St. Dymphna, the patron saint of the mentally ill, the psychiatrist started doling out my medicine. Because I refused to go to the hospital, I couldn't get a prescription for an amount that would last longer than three days, and I had to submit to periodic blood tests to show that the drug levels were stable. Though no one said so, I assumed these were safeguards to prevent me from stockpiling pills. Then Christmas came and everything changed. I was going home for three weeks. "I'll be gone the week you get back," the psychiatrist said when I told him, and without hesitation, he handed me a prescription for a month's supply. I finally had enough drugs to do it.

· 3 ·

"I never dream about my dad," my mother said. It was Christmas Eve, and we had just come home from Mary's. My father had already gone up to bed. "Christmas tomorrow," he had said when he bent down to kiss my mother. She was lying on the couch.

"Thanks for clearing that up," she said.

"Christmas tomorrow," he repeated, grinning, when he came over to kiss me. I was sitting on the loveseat, and as he bent down, I remembered a dream I'd had about him. He was in his robe, walking from person to person. When he got to me, he said, "It's Sunday today and Monday tomorrow. That might be fine for you, but all my trees need trimming," he said to someone else.

After I told her the dream, my mother said, "That doesn't sound like a dream, it sounds like real life with your father." And then she added, "I never dream about my dad."

"Do you dream about your mother?" I asked.

"Yeah, I do. A lot."

"Hmm," I said.

"Hmm what?" But before I could answer, she said, "I don't think it's because of the way he died. I dream about Sean. Anyway, I always gave my dad the benefit of the doubt. I think he deserves that. It could easily have been an accident."

"Mom, I was just thinking that I always have dreams about Dad but I never have any about you. That's all. We're the opposite of each other."

"Oh, well, I guess we are," she said, yawning.

Though it sounds as if she might be, my mother is not defensive about her father's death. There are two ways the story could be told, and my mother has always allowed for the possibility that his death was an accident, that his gun went off while he was cleaning it. It could be one way or the other, she always says, but she feels, as his daughter, that she owes him the benefit of the doubt.

"The benefit of the doubt" is the kind of phrase that confuses me, the kind my mother often uses. I remember the first time I heard it. My mother was giving Michael, Mary, and me the benefit of the doubt about something we had done. I was six or seven. It looked as if Michael and Mary knew what she meant, but I spent most of the afternoon trying to figure out what sounded to me like the oddest arrangement of words I'd ever heard. There's doubt, I told myself as I lay on the living room floor, staring at the ceiling, and it always appeared like a cartoon cloud in my mind, an airy, elusive, bubble-shaped substance. And then there's part of the doubt that's a benefit, I thought, but at that point the cloud disappeared. Okay, there's doubt. And I started over.

It took me years to realize that I was beginning with the wrong definition of doubt. It hadn't occurred to me that doubt could have a positive aspect. To me, doubt was Mary, hands on her hips, challenging some claim I had made. ("It's mine . . . you gave it to me . . . I had it first." "Uh, I *doubt* it," she would say.) Or my father, ending a punishment with "I *doubt* you'll be doing that again." Or any number of people doubting whatever they doubted that had to do with me.

"It's two things," Ellis would tell me, years later, after we had known each other for only a few months. She was exasperated by my inability to understand what she was saying. Ellis too was big on the benefit of the doubt. "Common idioms," she said, throwing up her hands and leaving the room. "It's two things," she said when she returned. "First, you've always been too literal. And second, you've never been able to understand absence. As a concept, I mean."

I didn't understand that either, but it didn't matter. Ellis had said something to me that was much more important: that I've always been and that I've never been able. It was these kinds of absolutes that I loved, words that made sense to me, that sank in and stayed. Just as my mother had done throughout my life, Ellis had defined me. She had definite ideas about who I was. To her, I had always been. I had never been able. Her words turned assessments into intimacies, extending time, taking it to a point before before and after after, long past the meeting and the parting, the coming and going, the excitement and sorrow, long past the good and the bad. You've always been. You've never been able. She might as well have said she loved me.

And she was right. I have always been too literal. I have never been able to understand . . . well, not absence really. It was Ellis's way of saying something else. These were my limitations: abstractions, concepts, ideas. Fantasies, Ellis added. Yes. It was the difference between my dreams and my mother's.

My mother got up and went to the kitchen.

"Syrup or no syrup?" she said.

"Lots of syrup," I told her.

She came back and handed me a bowl of peppermint ice cream, its pink color completely concealed in chocolate.

We were talking about dreams, recurring ones. In my mother's, Sean often appears as an animal or a bird or a disembodied voice in a dream that is nothing but color, in which the world, or whatever dimension our dreams are dreamt in, is bright white or yellow or the softest shade of blue, a pleasant, peaceful expanse, and against it, around it, within it, Sean's voice is saying something simple: It's okay, Mom; I'm okay, Mom; everything is good with me. Though he is usually nonhuman, he sometimes surprises her and appears as himself, a child, a boy, her young son, smiling.

But however he comes, whatever he is, she recognizes him. It was through her dreams that she knew he had gone to heaven. A few weeks after his funeral, she woke up one morning feeling too sad to get out of bed. Falling asleep again, she entered a golden forest, with trees taller than she had ever seen, ground softer than she had ever walked upon, piled thick with golden pine needles. Shafts of light led her through the forest to a clearing, where she looked down to see four white fuzzy caterpillars crawling near her feet. She wanted to pick them up and take them home with her, but she was afraid to touch them. She stood in the clearing, filled with awe and sadness and fear, watching as the four white caterpillars started moving slowly away. She reached out. She wanted to take them home to show us, but she couldn't bring herself to touch them, even though she knew in her heart that they were meant for her to have. And then she heard Sean's voice in the forest, in the sky, in the air all around her. "Don't worry, Mom. I'll carry them for you. I'll take them home." And when she woke, a deep sense of calmness and certainty had replaced the sadness that she felt. She was sure Sean was safe and watching over us. She was sure he was happy in heaven.

The first time she told me the dream, it was her fear of the four white caterpillars that intrigued me. How unlike my mother, I thought, to fear what she delighted in when we were children, all the caterpillars she'd caught and kept for us. But then I began to see that the four white caterpillars were us — Michael, Mary, Kelly, and me. Losing one child had made her question her right to the rest of us. She had once said as much to me. She was afraid of reaching out to us, of doing the wrong thing, and she was afraid, at the same time, that she would lose us.

In her next dream, she was walking in a forest and saw a bird on a low branch of a pine tree, sitting very still, as if it were waiting there for her. The bird was fat and fluffy, pale turquoise. She wanted to touch it, but something stopped her, and as she hesitated, the bird flew up, perched on her shoulder, and nuzzled her neck. It was Sean, she said.

The morning after the dream, she went to the grocery store, and as she walked across the parking lot, a fat, fluffy bluebird flew in front of her. Later, the bird appeared at the beauty shop. Shortly after, she saw him in the back yard, flying from the birdbath to the kitchen windowsill, where he watched her as she did the dishes.

"I see him a lot," she said.

"The bluebird?"

"Yeah. And Sean."

We were telling each other our dreams. I set my empty dish on the end table — "More?" my mother asked — and picked up the snow globe my parents had received a few weeks earlier from the Mercy Home for Boys. It was a large, heavy glass globe on a carved wooden base, a gift for all the money my mother and father had given in Sean's name. I shook it and watched the snow settle over the gold angel kneeling inside, its wings spread around two children. Someone from the Mercy Home had come all the way from Chicago to present the globe and thank my parents in person. They had been invited, as well, to a special dinner downtown — at the Marriott, my mother told me — to honor the Mercy Home's supporters in St. Louis, of whom my parents were among the most loyal and long-standing. My mother had called me afterward to describe the evening in detail, telling me how much it meant to her to meet the priest who ran the home. He hadn't been there the day I drove Sean's clothes to Chicago, but several weeks later he had written to thank her. With the letter, he had enclosed a receipt. In the years since, my mother had corresponded with him occasionally, and each time she wrote, she was touched by his attentive and caring response. The snow globe had been completely unexpected — a true gift, my mother called it. I shook it and watched the snow swirl up again, a flurry of sparkling flecks suspended in fluid, and as I watched the snow cover the angel, changing its wings from gold to white, I amended my dreams so that Sean was never naked.

In my dreams, Sean was always himself. Sometimes he was a small child, two or three. Sometimes he looked as he had in grade school. But in my most recent dreams, he was at his oldest, as if I had dreamt him through his whole life. I was worried that this meant he would disappear soon, leaving my dreams as devoid of him as my mother's were of her father. I hoped this wouldn't happen. I wished he would assume a younger age again, one that would allow us to start over. It was almost a decade since he died. What could I do to make sure he stayed with me? Sometimes before I fell asleep, I looked at pictures of him as a baby or a young boy. But he resisted my desires. He remained forever fifteen.

"That's an odd one," my mother said when I finished telling her about the dream in which I carried Sean through a card shop. I was holding him as if he were a baby. But he was a teenager. His legs dangled over one of my arms, his head lay against the other, and the rest of his body hung heavily in between. It was hard to walk with him. We moved slowly through the store, up one aisle and down another. Every few feet he would signal for me to stop, and he would reach out for a card, read it, shake his head, and return it to the rack. He rejected one after another. He was naked, of course, and sick, pale and weak and lacking the strength to walk on his own. What we were doing did not seem unusual. It was as if I carried him wherever we went, but in this dream I could see, as I carried him through the card shop, that he was dying.

I told my mother a few of the other dreams I'd had about him in the months preceding Christmas, making each dream sound more hopeful than it had been, with Sean appearing happier and more alive, dressed and ready to walk on his own in the world. Down to the details of my dreams, it had become my habit never to describe my life fully to my mother — or to anyone else, for that matter. Not to Mary, Michael, or Kelly. Not to my father. Not to my friends. It was a lonely way to live. I wanted to disappear. I wanted to die without having to do it.

"We better go up," my mother said. "Mass will be crowded tomorrow for Christmas. We'll need to get there early if we want a seat."

It was close to two in the morning. I wasn't tired; I knew I wouldn't sleep.

Hahhh, hahhh, my mother sighed as she started rocking

back and forth on the couch, gaining the momentum she needed to stand up.

I kissed her goodnight and put the dishes in the sink. Upstairs, I lay in bed and listened as she made her way from room to room with her dust rag. When she came in, I closed my eyes, and she played along, letting me act as if I were asleep.

I lay awake a long time. I heard my father get up and go downstairs. It must have been morning, or my father's version of morning — four-forty-five, I thought. It was Christmas. Most of the holiday was behind us. Over the years, Christmas Eve had become the high point, when we ate the bigger and fancier of the two days' dinners and gave out all the gifts. This we did at Mary's. She had the largest house. What lay ahead was a simple dinner my mother would make, with everyone bringing something over to add to it. This was the way it was at my parents' house on Christmas. We would go to Mass, then spend the rest of the day slowly setting things out for the evening. I would be sent to the 7-Eleven sometimes, the only store that was open, for milk or film or some other last-minute thing that was missing. That was the newest oddity about Christmas: that something would be missing, a common ingredient overlooked, my mother discovering it mid-recipe — a can of clear broth or condensed milk, a cup of seasoned bread crumbs. It was things like this — buying a Big Gulp at 7-Eleven on Christmas Day and drinking it while I drove home — that I'd never foreseen for myself. Or that on the way back from the store I would think of stopping at the cemetery instead of going straight home, that there would be a reason to stop there, that on what had once been a busy day like Christmas, there would be all the time in the world to do it.

I must have fallen asleep. My mother was waking me for Mass.

"You better get up if you're coming," she said.

My mother was a Eucharistic minister at Christmas Mass, one of a number of lay people who helped the priest give out communion. It felt strange receiving communion from my mother, but she was the only one serving the section of pews where my father and I were sitting.

"The Body of Christ," she said as she placed the wafer on my tongue. "The Body of Christ, Tom," I heard her say to my father, personalizing, as if by habit, what she put in his mouth.

The Body of Christ, she said, and she believed it, giving communion to one parishioner after another. Back in the pew, I dislodged the host from where it had stuck to the roof of my mouth and swallowed. I never took the wine that went with it. ("The *blood*," Allison corrected, fresh from her First Communion. "The blood," she said, "is *really, really* good.")

As I watched my mother, I thought about the vision she had had the week before Sean died — all the boys from Sean's basketball team appearing at the altar, holding a casket that turned out to be his. What did she believe? Why did she believe it? "The death of a child is the greatest reason to doubt the existence of God," I read once. It was Dostoyevsky who wrote it; I don't remember where. But they are among the few words I've read that I've never forgotten. After Sean's death, my mother's faith never faltered. If anything, it flourished. My father seemed to remain firm in his beliefs as well, though he professed his faith more privately than my mother.

I remembered how excited she was when each of us received our sacraments, when we made our first confession, our First Communion, our confirmation, when Sean and Kelly were baptized. Her excitement extended now to Mary's kids and Kelly's. "Dear God, I am hardly sorry," Jesse said, practicing the prayer he needed to learn for his first confession. "*Heartily*, Jesse. Heartily. *Very sorry*," she corrected, charmed by her grandson as he prepared for this, his second sacrament.

Around the same time, she received another sacrament herself, the sacrament of extreme unction. She hadn't been sick; receiving the last rites was precautionary on her part. She intended to do it again, often, not just when she was sick or death seemed near. She would have herself regularly reanointed, a kind of insurance policy, she said, something that might improve her chances of seeing Sean again.

Partly she feared we would forget. If my father died before she did, would we remember to call a priest? It wasn't an idle worry. We had failed her once already — when she suffered the illness that caused her to lose her memory.

She had been home from the hospital for a few weeks, and in an attempt to reconstruct what she couldn't remember, she asked me to tell her again what had happened.

"Mary and I found you lying on your bedroom floor in the late afternoon. It looked as if you had collapsed there. You were delirious and weak; you didn't know where you were, and you kept passing out and coming to again. We called the doctor, and he told us to give you hot tea, put you to bed, and call him back in the morning if you weren't feeling any better. We said we thought something was really wrong, but he repeated that you probably just needed some rest. Then you started having convulsions, so we called an ambulance and phoned the doctor again, and he told us which hospital we should go to. He'd meet us there. You were completely unconscious by the time the ambulance arrived."

"Why can't I remember it?" she asked.

"I don't know," I said. "You were really sick."

"What made me that way?"

"You lost a lot of the potassium in your body. Almost all of it," I answered.

"Who found me?" And we would go through it again, day after day, the account never changing, her response always the same, until one day she surprised me and said something different.

"I guess I was given the last rites. That means I've received all the sacraments available to me." The thought of it seemed to please her. I didn't know what to say. In all the commotion, it hadn't occurred to Mary or me to call a priest.

I didn't tell her all the details. I didn't mention, for instance, that during those days I hadn't given her the benefit of the doubt. When her doctor told us that she had taken an overdose, I believed him. She had been depressed. "Attempted suicide," he wrote on her chart.

"That *can't* be true," Mary insisted, repeating what my mother had said when we found her on the floor. "It's not like my father; make sure they know it's not like my father."

Mary was right. What the doctor wrote on my mother's chart wasn't true.

"Go home and count the pills left in her prescription bottle. They pumped her stomach, but there weren't any pills in it," a young intern finally came forward and told my father.

My mother's doctor had prescribed medication that was known to deplete the body's store of potassium, but he had neglected to tell her to start taking potassium supplements or to add potassium-rich foods to her diet. "Attempted suicide," he wrote on the chart. I believed him, but Mary didn't. Taking to heart the words we had heard my mother say — "Tell them it's not like my father; make sure they know it's not like my father" — she stood up for her, while I remained silent.

One night when Mary and I were driving home from the hospital, I started crying. My mother was conscious by then, and though the doctors were confident she would regain her physical functions, they weren't sure how much of her memory she would recover. It was Mary, of course, who comforted me. As a diversion, she decided I should learn how to drive. We were in my mother's car. I was fifteen; Mary was seventeen. In a few weeks, I would be a sophomore in high school and Mary would be a senior. The following summer, after she graduated, she would become the manager of the card shop where she worked at the mall. But that was a future far ahead of us. That night, we were facing a future that was more immediate, one that centered on the fact that my mother's mind had become frighteningly feeble.

"You'll be sixteen before you know it," Mary said, trading seats with me when we were a few blocks from home. My mother's car was an automatic, so it didn't take much skill to move it forward. I went up the hill and thought I had the hang of it, but at the first turn, I lost control and jumped the curb. By the time I hit the brakes, we had run over a row of bushes and were in the middle of someone's yard.

"Shit," Mary said, and she reached over and shut off the headlights. "Trade places," she whispered, and she took the wheel and drove the car back into the street, leaving the headlights off until we were around the corner and halfway down the block.

The next day, as we drove past the house on our way to the hospital, we saw that half the hedge had been uprooted. A long trail of tire tracks divided the lawn.

"Oh, my God," Mary said. We slowed down but didn't stop. We could see now, in daylight, that the lawn had been recently dug up and replanted. Plugs of zoysia dotted the soil like little islands of possibility.

"Jeez," I said. In the tire tracks, the zoysia plugs had been pushed completely underground. My first attempt at driving had ruined a big investment. I had run over the kind of yard my father had always hoped to have for our house, the kind we could never afford. I imagined how my father would feel if this had happened to him. He was the only man I ever ached for. It made me sick to see what I had done.

Mary and I kept on going. We didn't even ask each other whether we should stop. I was thinking how this would always be the story of the first time I drove, a story I would be too ashamed to tell, so different from the story my mother always told us about her first time driving. She had taken her father's car out one Saturday afternoon when her parents were away. Like me, she was a few months shy of sixteen, but she'd watched her father so often that she was sure she could do it. She picked up some of her friends and drove to Heine Meine's Baseball Park. There was a road in the park that circled a big field full of baseball diamonds. She dropped off her friends and drove around the road. Every time she passed, her friends called to her to park the car and come watch the game. A boy she liked was playing. But my mother insisted she wanted to practice driving. She wasn't sure when she would be able to get the car again, she said. So she circled Heine Meine's until the game was over. Why hadn't she pulled into the parking lot and joined her friends? She couldn't remember how to put the car in reverse and was afraid of having to back out of the lot later. To avoid embarrassment, she decided the best thing to do was to keep moving forward, around and around the drive at Heine Meine's. It was one of my mother's favorite stories. It was one of mine too.

Mary and I were on the highway now, nearing the hospital. We hadn't spoken at all. We hadn't stopped at the man's house, and we hadn't spoken. From now on we would avoid passing the house, steering clear of our shame night after night. My mother's only embarrassment had been not knowing how to drive in reverse. Why was it that my parents' stories were always so simple? Why were they so often filled with odd names like Heine Meine? (When my father was a child, a man named Heine Butz lived upstairs from his family and kept cold cuts in their icebox. "Don't eat Heine Butz's baloney," my father's mother would call out whenever she heard the kids in the kitchen.) Why were their stories so innocent and unforgettable? I wanted to say something to Mary about Heine Meine, something to break the silence, but I didn't. We were almost at the hospital. Maybe she was even thinking about the same thing. Maybe she was remembering the story about Heine Meine's too and wondering, as I was, whether it would be among the many memories my mother had suddenly lost, a story that in time we would have to tell her.

When we got to the hospital, my mother wasn't there. She had been transferred to the psychiatric ward that morning. The day before, my father had confronted the doctor about the cover-up he had manufactured, falsifying my mother's chart at every turn, lying about the cause of her condition. This was the doctor's response to my father's accusations: he had gotten my mother to sign herself in. Her signature was shown to us. Lois Schneider, it said, in a faint, shaky semblance of my mother's handwriting. She had forgotten she was married.

She didn't realize anything was wrong; she thought she'd been moved because her condition was improving. "They have craft classes here," she told us.

The next day my father signed papers declaring my mother mentally incapable of making medical decisions. When he refused to authorize the shock treatments the doctor recommended, the hospital said it had no choice but to discharge her. The day my father brought her home, she was still weak and disoriented, clutching the white yarn octopus she had managed to make.

She was forty-one when that happened. She was fifty when Sean died. And now, approaching sixty, she was giving out communion at Christmas, standing at the altar looking happy, as if she had never suffered such hardships. "The Body of Christ," and she believed it. "Peace be with you," and she believed it. She believed in Jesus and God and the Church. She believed in being Catholic.

"Peace be with you," my father had said, kissing my cheek before communion. "Peace be with you," said my mother. It was a formal part of the Mass. Peace to one another, neighbors, strangers, family, friends, each extending a hand to say it. Peace to everyone around us. "Peace be with you. Merry Christmas. I love you," my father had said, the same to me as to my mother, kissing each of us in church. "Peace be with you. Merry Christmas. I love you too, Tom," from my mother.

It was this part of the Mass — where people took each other's hands and said *Peace* — that made me think of Sean. This part, and every time the bells were rung on the altar. Whenever I heard them, I thought of the boy who rang the bells too soon. His mother had approached me at Sean's funeral. She was a member of the parish, and though she knew none of us, she wanted to tell us how Sean had helped her young son learn to be an altar boy. Her son was nervous. He would often ring the bells at the wrong time, and Sean, serving Mass with him, would reach out and take his hand whenever he saw him starting to ring them too soon. "He was so kind," the woman said, her eyes filling with tears. "He was so gentle," and she took my hand and held it, showing me how Sean had quietly held her son's hand over the altar bells until it was exactly the right time to ring them.

The church was crowded, and the Christmas communion service took much longer than a regular Sunday Mass. While my mother stood at the altar handing out the hosts, I listened to my father sing. My parents were people of profound faith. In all things, alone and together, they tended to give God the benefit of the doubt. At home, my father had begun to read the Bible. "That book's not for everybody," my mother had said. She preferred the little booklets of spiritual messages she received each month in the mail. She had been subscribing to the *Daily Word* for over a decade. Sometimes there were short essays at the beginning of the booklet by, as my mother put it, "well-known people who have good relationships with God." She enjoyed reading their thoughts, learning how God played an important part in the life of Bob Barker, Phyllis Diller, Marian Wright Edelman, and Famous Amos, but mostly she liked the daily meditations, based on "manageable" quotations from the Bible. If a meditation moved her in a particular way, she tore out the page and stuck it in her prayer book. "I Am Triumphant!" was the message on the day Sean died. "I am Christ in you, the hope of glory."

Earlier that year, my mother had sent me a subscription to the *Daily Word*; it started coming in the summer. It had been a bad year. My depression had grown greater than all the skills I had developed to disguise it. There was no hiding it now. My mother grabbed at whatever she thought might help, making sure that she shared with me anything that had had even the slightest beneficial effect on her life, on her ability to deal with her depression. On the days I read the *Daily Word*, I could understand its appeal. There was something simple and inspiring about the messages, something that wasn't difficult to comprehend or remember. Something that didn't make you feel any worse.

"Go in peace," the priest said at the end of the Mass. My mother was back in the pew, standing beside my father. I could see that she was at peace. It was as if all the peace professed at Mass each Sunday had settled inside her. But it was more than that. It was the way she had withstood the illnesses, the de-

pressions, the deaths, the years of economic uncertainty. It was
the way she kept looking ahead and leaving the bad behind
her. It was as if my mother's first day of driving had become a
metaphor for her whole life. She had mastered only the mechanics for moving forward.

That evening, everyone came for Christmas dinner. My mother
decided to change the usual seating arrangements by having all
the grandchildren eat in the dining room. Michael, Kelly, and I
joined them, and everyone else ate at what had always been
the kids' table, in the kitchen.

"Are you sure you want to do this, Mom?" Mary asked before we sat down.

"Oh, why not?" my mother said. "They're always asking to
eat in the dining room."

Allison sat at the head of the table and said grace. All of her
prayers ended the same way. "And thank you for Shamu," she
said. She had been expressing gratitude for the whale every
day since seeing him at Sea World the previous summer.

"What is it about Shamu that you like so much?" Michael
asked her.

"Oh, Michael!" she said, with all the intensity of a seven-
year-old. "He's the most beautiful whale I've ever seen!"

We all laughed.

"He is!" she shrieked.

"Al, he's the only whale you've ever seen," Jesse told her.

"So?" she said.

Beneath her seat at the table was the box turtle she had received for Christmas. "It's a female," she told us before dinner,
pointing out all the indicators of the turtle's gender. At the
pet store, her name was moving up the list for a male. The
fact that she'd gotten a female for Christmas — "from Santa
Claus," she said with a knowing grin — freed up half of her
First Communion money. She was thinking about buying a
parakeet, I heard her tell my mother.

"What about a pair of little finches?" my mother suggested,
and they went to the basement together, Jesse too, to find the

cage and the paraphernalia for the finches Sean had bought my mother one year for her birthday. When they came up, Allison was carrying the cage and Jesse had an armful of basketball trophies he had found. Some were mine, some were Michael's, and some were Sean's.

"Can I have these?" he asked. He wanted to put them on his dresser, he said, until he got some of his own.

"Sure," Michael said. I told him he could have mine too.

"What about Sean's?"

"You can have them, but let me keep this one," my father said, removing the only trophy that wasn't for basketball. A gold runner stood on top.

"That's Sean's cross-country trophy," Jesse informed my father. Mary's kids all spoke about Sean as if they'd known him. I wasn't sure whether this was my mother's influence or Mary's. Years later, when they were teenagers and Mary thought she should tell them the truth about Sean's death, Jesse cried and said he thought it was the saddest thing he'd ever heard. "Why did he do it, Mom? Why did he have to do it?"

Aside from the change in seating arrangements, it was our usual Christmas, and after everyone left, my mother and I resumed our routine. But I was too tired to talk, and soon after we finished our ice cream, I told her I was going to bed.

"Ohh! Really? Okay," she said. I could see she was disappointed.

I went to bed early again the next night, and the day after that we went to Mary's lake house in the Ozarks. We took walks; we shopped at the outlet malls; we visited relatives and watched movies. Every morning Allison and my mother walked along the lakeshore looking for turtles, even though there wasn't much hope of seeing one until spring. "You never know," my mother said. One morning she was sure she saw a blue heron near the boathouse. "I didn't think they stayed here in winter," she told Dan. "What happens when a blue heron eats a red herring?" my father asked, but he couldn't come up with an answer that worked. "There's something funny in there," he insisted, asking the question over and over until my mother finally told him to give it a rest.

Each evening before the kids went to bed we played games — Pictionary and Scrabble and Scattergories and Monopoly and Life and even Uno, which the kids couldn't stand. "*Bor*-ing," Sarah said whenever she was down to her last card and could call out "Uno." "*Bor*-ing," she'd announce repeatedly until she finally won and made us promise never to bother her again with that game. "Yeah," Jesse said, "it's the week of bored games. Get it?"

It had been Mary's idea to go to the lake. We'd never been there together in winter, and she thought it might make me feel better. It was for me, my mother said, that Mary seemed to have the most compassion. It was true. "It's you," Mary told me when we were younger. "It's you," she had said after telling me how my mother studied our faces as we slept, trying to determine which of us would inherit the depression that ran through our family on both sides. Even though that night I hated her taunts, denying them as we lay in the dark, I always knew that if it *was* true, if I did turn out to be the one, she would be there for me, her compassion growing with each episode. My mother maintained that it was Sean who was saving me, watching over me, making sure I stayed alive. It wasn't. It was Mary. It was Mary, from however far away, covering me with her wealth and her houses and her family, spreading her successes over me, sharing everything she had. It wasn't anything spiritual or mysterious. It was something more material and immediate. It was riding on a Jet Ski with Jesse. It was hearing Sarah sing or listening to the stories Allison made up. (Once upon a time there was a beautiful whale who lived in the creek behind our house, and one day God came by and asked the whale to teach him how to swim. The whale tried hard to teach God, but God was a terrible swimmer. "You're the worst swimmer I've ever seen," the whale said. And God cried and cried, and the whale said to God, "Come on now. Snap out of it!" And God stopped crying and started to practice swimming, and he got a little better, but not very much.)

Or sometimes it was Michael, calling with his latest assortment of curious facts, asking whether I knew that Hawaii was home to the largest cattle ranch in the United States and supplying me with the acreage to substantiate this claim. Or it was Kelly, or, in time, Kelly's kids. It was Stephanie's first attempts at telling jokes. "Knock knock. Who's there? No one. It's just me. I'm telling a joke." Or Nicholas, at two, stomping around the house in his beloved snow boots, refusing to wear anything but the blue rubber boots and a diaper, the sight of his little body its own kind of blessing. Or it was my father. Or my mother. It was each of them. It was all of them.

And Mary was right. By the time the week was over, I felt better.

A few days after we returned from the lake, a boy shot himself in a small town outside St. Louis, and we woke up to find the story on the front page of the *Post-Dispatch*. The boy was a high school basketball player, a starting guard for one of the state's highest-ranked teams. The story carried a color photograph of his funeral. It was held in the high school gymnasium, on the basketball court, and the picture showed the boy laid out in his coffin right below the basket. A basketball had been placed in the coffin, and I couldn't help noticing that it was positioned exactly below the net, as if someone had just made a perfect shot and the ball had landed on the boy's body, coming to rest at his waist. Would they bury the basketball with the boy? I wondered.

The story opened:

> Hundreds of townspeople filed silently across the high school basketball court Friday under the darkened scoreboard that bore his name and past the casket that bore his body. Why, they wanted to know, would a starting guard of the state's third-ranked Class A team simply attend practice as usual New Year's morning, and then — still in his warm-ups — blast himself in the chest with a shot gun at home that afternoon? The handsome 18-year-old honor student left them no answer.

The story went on to report that the basketball team had hoped to recapture the Class A state title it had won the previous year, but two days before the boy shot himself, the team suffered a two-point loss in the championships. "Officials said they did not believe that should have been enough of a disappointment to explain what happened," but a prominent St. Louis psychiatrist who had studied suicide for thirty-nine years said, "Such a trivial upset could definitely trigger suicide in someone already depressed."

None of us spoke about the article, and after we had all read it, my mother cut it out and put it in her prayer book.

I couldn't help thinking about basketball and the part it seemed to play in the boy's death and in Sean's. After Sean died, several weeks passed before we learned what had really happened. The news came to us through a friend of Mary's whose brother had been on Sean's basketball team — not the St. Thomas Apostle grade school team he had helped lead to the league finals, but the freshman team he had been picked for when he moved to Florissant Junior High and started public school, as Mary and I had done before him, as Kelly would do a few years later, each of us ending our Catholic education after eighth grade.

That year, he was no longer the star forward he had always been, playing for a small Catholic school. Instead, he played second string and spent every game on the bench. Mostly he was bored, he told me just after the season started. He had found the sport he loved in cross-country, he said, and he felt that basketball was behind him.

"What should I do?" he had asked. "I shouldn't even have tried out."

"Quit," I told him.

"I can't just quit."

"Why not?" I asked, but he didn't answer.

Partly it was his competitive nature, but it was something else as well. It was perfectionism combined with the pressures of adolescence. I had played basketball through grade school and part of high school too, and hated every minute. But until

I failed to make the team in my junior year — intentionally playing badly during tryouts — I hadn't quit either. Maybe it had been the same way for Michael. After his sophomore year, he let basketball slip out of his life quietly and uneventfully, as I had.

Though Sean couldn't quit, he already seemed ahead of Michael and me. He wouldn't suffer through a sophomore year as we had. After ninth-grade basketball ended, it would only be cross-country, track, and tennis for him, he had told me. He couldn't wait for spring, when track would start at school and the weather would be right for playing tennis.

But first there would be that final freshman season. On the day he died, Florissant Junior High was winning by a large margin. With little time left on the clock, the coach felt it safe to send in the second string. In those final few minutes, Sean made his first shot of the season. He must have felt so alive, so exhilarated, to come off the bench and score. The crowd went wild. Everyone in the auditorium began to scream and clap, shouting out his name and his number. Later, the coach singled him out in the locker room. Never in his career had he seen a shot like Sean's, he said.

That night, at home, he didn't mention it. At dinner, he never said a word about the game. After he finished doing the dishes, he went upstairs, made a timid attempt at slitting his wrists, then opted to swallow a fistful of the pills that keep my father's heart beating in a regular rhythm. "I hate basketball and am no good at it," he wrote in the letter he left. In the few minutes that he had been sent in to play, he had taken the ball down court the wrong way. The shot he made was to the other team's basket.

A "trivial upset" was how the psychiatrist had put it in the morning paper, commenting on the suicide of the boy whose team lost its chance at the state title. I thought about the picture, the boy's body laid out under the basket, the scoreboard bearing his name. Mostly, I thought about the basketball that had been placed in the casket. Would they bury the ball with him? I had slipped a buckeye into the breast pocket of Sean's suit before he was buried. We picked up buckeyes in the back yard every fall when they fell, believing what my mother told us — that buckeyes brought good luck. In the beginning, whenever I thought of him in the ground, I saw only that single buckeye, the rich, brown, shiny roundness of it, like a chestnut suspended in darkness.

I pictured the boy's basketball slowly deflating beneath the ground, losing its air in a long, slow sigh, as if it were letting out its last breath, and I wondered whether deflation was always the first stage of decay. Houses settle, my father had told us. Bodies settled too, I knew. I imagined the basketball collapsing into itself, changing into a concave bowl. It was a sad and empty image, unlike the one I had of Sean — a buckeye tree taking root in his breast pocket, branching out above his heart, bearing fruit and dropping its seeds every fall, over his body, over his bones, like a little good-luck garden underground, a buckeye tree, bonsai-size, still breathing.

I was still thinking about the boy when my mother and I sat on the sofa that night eating ice cream. It had been a shock to see the story in the morning paper. My mother must have been thinking about it too. She was glad, she said, that Sean hadn't shot himself.

Was it because her father had shot himself? I wondered.

"No! No, it's not that. It just seems . . . I don't know, so much more awful somehow."

I didn't say anything.

"But then, if he had, we would have heard it," she said. "Maybe that would have made a difference."

"I doubt it. People usually shoot themselves right in the head or the chest," I said. "That's hard to survive at close range."

My mother looked at me, and I could see a kind of horror on her face, as if I had provided her with unwanted information. It registered like a shockwave under her skin. I wasn't sure whether she was reacting to a memory of her father — had he shot himself in the head? the chest? — or to the matter-of-fact way I had spoken.

I had only recently begun to think of committing suicide in ways I had never before considered. It was as if, over time, depression eroded the natural fears that served as my defenses, leaving my mind open to a wider range of options. There was a time when I couldn't have imagined shooting myself or jumping from a high floor of a building. For most of my life, I couldn't entertain the slightest thought of a suicide that required more involvement on my part than opening a bottle of pills and swallowing them. The bodily pain, the violence of other methods, made me shudder, in the way I shuddered when I thought about Sean's wrists. But I had lost all that. I had lost the necessary balance between fear and boldness — no, between fear and no fear. Was this how people died of depression, not suddenly but slowly? Maybe our drastic, desperate endings, our suicides, only appear to be sudden. Maybe they are merely last-minute acts that mask a slow degeneration over time, as if the very will within our cells, our instinct to live, is gradually extinguished, leaving us physically whole but hollow.

"Do you *own* a gun?" Ellis had asked when I phoned to tell her of this new development in the history of my depression. It was the first thing she said, and it touched me in the deepest way — the calmest and most practical question, Ellis going straight to the heart of it.

"Are there any ways of doing it that you're still afraid of, that you'd never attempt?" she asked.

"Yeah. Hanging myself."

"You're afraid of it?"

"No. I'd never attempt it. I'm too heavy." I was serious; it wouldn't work. Unless I did it outdoors from a big sturdy tree, there was nothing I could hang a rope around that would hold my weight. It seemed to me that only normal-size people — maybe only thin people — could successfully hang themselves in the shower or from a light fixture or a wooden beam. For me, it would be not only futile but embarrassing. I was serious, and Ellis tried to listen with her usual empathy, but she couldn't help it; she started to laugh and couldn't stop.

"Why are you grinning?" my mother asked. "What are you thinking about?"

"Oh, nothing really."

"It must be something."

"It isn't," I said, and I started to laugh.

"All right, then, leave me out of it," she said.

She was lying on the couch, and she turned on her side to look at me. I could tell from her expression that she'd be laughing soon herself. I have always had that effect on her. ("What is it about you with Mom?" Mary asked me once. "You just look at her and she starts laughing. She spends half the time in the bathroom when you're home.")

I tried to say something, but I couldn't get the words out through the laughter.

My mother was smiling now, and she swung her legs off the couch to sit up, sighing — *hahhh, hahhh* — as she did it.

"*Hahhh, hahhh,*" I said, and then she started laughing her silent laugh, her face turning red, her upper body shaking, no noise coming out, not even a wheezing sound. Soon she would be laughing so hard she'd put her hand on her chest and roll back on the couch.

I was laughing about myself — about being self-conscious even when it came to suicide — but then the whole absurdity of it hit me. My mother was glad Sean hadn't shot himself. Wasn't I glad too? she'd wondered. "Yes," I said. "It would have been so much worse if he'd shot . . . if he'd shot himself" — the words came out in gulps through the laughter — "because then he'd *really* be dead."

"It's not funny," my mother was trying to tell me, putting her hand up to say something and then letting it fall back onto her chest. She laughed and laughed, her face turning redder and redder. When I told her that I was too heavy to hang myself, that was it. She rocked herself up from the couch and ran to the bathroom.

"Ohhh," she said when she came back, the last red splotches of laughter still on her face. "Ohhh, that's it for me," she said, collapsing on the couch, and we laughed a little more, letting the joy of it slowly settle.

A few nights later, before returning to New York, I had the dream I'd hoped I would never have. I was at a picnic with lots of people. It seemed as if everyone I knew was there — my entire family and all my friends. I had the feeling that no one was missing. I was sitting at a picnic table under a pavilion, playing a card game with my grandmother. Kings in the Corner, it was called. "I like my weenies burned," my grandmother kept saying to a man standing at the barbecue pit. Each time it embarrassed me, as it had in real life. "Do you kids want some weenies?" she'd ask us whenever we spent the night at her house. In the dream, the man at the barbecue pit winked at me when she said it. Sometimes the man was my father; sometimes it was Michael or an uncle of ours. "Hot dogs, Grandma, hot dogs," Sean said with a laugh, straddling the bench beside her. He was wearing jeans and a T-shirt and was bouncing a tennis ball on the bench. "Stop that, Sean," my grandmother said, and he ran off to hit the ball against a wall.

My mother and Mary sat down, and my grandmother changed the game to gin rummy. Next to the pavilion, there was a pool, where my mother was teaching Kelly how to swim. At the same time, my mother was still at the table playing cards with us. Though Sean looked fifteen in the dream, Kelly was only two or three. My mother was helping her float. "Can you let her go yet? Can she float?" I kept asking. Sean came back. "Can I play?" he asked, but before we dealt him in he was gone again, and a little later I saw him in the pool, swimming laps by himself.

The day went on with all kinds of activities, but we kept playing cards, people getting in the game and going out, and suddenly I realized that I hadn't seen Sean in a while. I put down my cards and went looking for him. I looked by the tennis courts and the playground. I looked among the people playing volleyball. I went to the pool and got in. I thought I saw him swimming near me, under water, the flash of his body going by, but when I swam to him, it turned out to be someone else. I got out of the pool and searched the picnic grounds, but I couldn't find him. Finally, I asked a friend whether she had seen him. As I spoke, I wondered why I was asking her. She was a friend from New York who had never known him. "Sean?" She smiled, and her face lit up. "He was here," she said, "but he left early."

"He was here, but he left early," I said to my mother, telling her the dream the next day. It was such a simple statement. "He was here, but he left early." I started crying as I told her, though it wasn't really sadness that made me cry. A flood of relief flowed through me, a long-awaited ending. I had a strong sense that I wouldn't be seeing him in my dreams again — or that if I saw him, it would not be soon — but instead of feeling sad, I felt strangely peaceful. In the dream, we had a happy time together, and then he was gone. Still, I was crying. He left before the rest of us. He never came back to say goodbye.

"It's a good dream," my mother said. "It's gone on too long for you. I was getting worried. It's a good dream," she said, kissing my head.

What I suspected was true. It would be another six years before I saw him, and when he appeared, he was fifteen, the age he had been when he died. In the dream, he was walking back and forth, fully clothed, at the end of Mary's front yard, with a baby in his arms. Mary and I were on the front porch. "Where are you taking the baby?" Mary shouted to him. "It's okay. I'm holding on to her," he yelled back. I assumed the baby was Sarah. In the dream, we were at Mary's first house, the one Sean had ridden his bike to every day after school, his heart set on seeing Sarah. It was Mary's house, but instead of a street at the end of the yard, there was a four-lane highway. Sean walked back and forth beside it, cradling Sarah. Cars were speeding past. "Where are you taking the baby?" Mary shouted. "It's okay. I'm holding on to her," he yelled back.

In the six years between my last dream of Sean and the dream of him holding the baby, Sarah had turned sixteen. The rescue workers didn't know how she survived the crash that killed her best friend. They were in a sports car, on their way to a high school hockey game with two other girls — it was the day after I had the dream — and a large van ran through an intersection into the driver's side. "God. Thank God she was on the passenger's side," I said after Mary told me the details. "She wasn't," Mary said. "She was sitting behind the driver. They don't know how she survived." Sarah walked away from the accident. Physically, she had hardly been hurt.

It would be six years between the dreams, and yes, he would go away again, but that day when I was still at home in St. Louis for Christmas, in the year that I came too close, my mother held me in her arms, and I knew the worst was over. "It's a good dream," she said, kissing my head. "It's the dream you had to have."

The next night, back in New York, I unpacked my suitcase and found a tiny book about butterflies that my mother had slipped inside. A note attached to it said, "I got you this book because it reminded me of Sean. One year for my birthday, he bought me a card with butterflies on it and gave it to me while I was still in bed. When I went downstairs, I found he had left a jar full of butterflies waiting for me on the kitchen window-sill. He had a special talent for giving cards and gifts. You have a special talent that way too, and I hope you never lose it. You are the best gift I get each Christmas. Every year, I feel happy when you're home."

That night, I lay in bed thinking about the future, about all the IMAX movies my mother would see before I came home the following Christmas, about how many babies Allison's box turtles would breed. "I'm next in line on the list for males!" she'd phoned the night before to tell me.

I thought about my mother. Somewhere in St. Louis a man was wearing her clothes. (In Chicago, was there a boy still wearing Sean's?) In the year to come she would surprise us all by taking a trip to Alaska. In the pictures my father took, she stands on top of a glacier, looking sure-footed. Maybe she would see Antarctica after all. Maybe I was wrong. Maybe there was room there for a woman who moved slowly and was afraid of ice. I wondered how it felt to live life the way my mother lived it. I wondered what it was like not to worry about who you were or how you looked.

It was after midnight. I saw snow starting to fall, and I got up and stood at my window. A fine white layer covered the street, the sidewalk, and the rooftops of all the buildings below me, and I could tell from the way it fell that it was a dry snow, a dry white radiance in which each tiny crystal lay loosely and lightly upon the others. Somehow it satisfied me that I understood this about snow, that I could discern its qualities from a distance, that I knew what it felt like from far away. I remembered the story my mother used to tell me about the snowstorm that shut down St. Louis on the night I was born, how frightened she and my father were as they drove through it, trying to reach the hospital. It was not this dry, delicate snow that fell the night I was born, but heavier, wetter snow, snow turning quickly to ice, the same snow that fell the night Sean died. I could still remember the hard wetness of it hitting Michael's windshield as we drove in silence, moving so slowly, trying hard to get home. I could still hear the sound of that snow against the sound of Michael screaming, hours later, standing in the room he once shared with Sean, screaming and crying as the snow fell hard outside our house. It was a wet, relentless snow that fell the night Sean died, not this dry, delicate snow that I watched falling silently outside my window now, and as I stood remembering it all these years later, I knew my mother was right. Like Michael, I should have started screaming sooner.

I watched the snow for a long while, and as I watched, I thought about the jar of butterflies Sean gave my mother for her birthday. He had inherited from her an affinity for animals and insects, for all forms of life. I imagined them at the end of the day releasing the butterflies in the back yard, watching them fly off into the darkness. That was an August night, long ago, in the suburbs of St. Louis. How old was he? I wondered. Which of the fifteen years of his life belonged to that birthday?

I watched the snow falling in New York now, all these years after, and I wondered how my mother — how both my parents — had borne the loss of him so well. I didn't understand. I would never understand. But I knew that somehow they had borne the loss of him for each of us, and by doing so, by grieving with such selflessness and grace, they had given us their greatest gift.

The snow fell, and I thought about butterflies, what kind of magic it would be to see one in winter. I thought about white fuzzy caterpillars in golden forests and bluebirds that followed my mother to the grocery store, the beauty shop, and back, and I wondered what my mother would see if she were standing at the window with me. Would I see it too? I wasn't sure what I believed — God, angels, miracles, bluebirds that might be my brother — but I still hoped that somewhere, someday, I'd see him.

Miles away, my mother was finishing off the ice cream without me, my father already asleep. Soon she would make her way through the house, dusting in the dark, and I knew, that night, more than most, she would miss me. That night, more than most, I hoped she knew I missed her too.

The Tender Land

One day in December, in the month before he died, Sean and I made a deal. "If you buy me a record for Christmas," he said, "I'll buy you a book." We already had years of gifts between us. He was fifteen, I was twenty-four, and we had been born into a family in which gift-giving is a dominant gene, a congenital characteristic that manifests itself in all of us as soon as we are old enough to save money and to spend it.

Sean's proposition was a bold one. It broke the rules of exchange, innate and unspoken, that governed gift-giving in our family. A gift is not a gift if the giver knows, before giving it, that it is exactly what the person who will receive it has wanted all along. That is another transaction. That is a purchase, not a present. A gift is a gift if you search for it, ponder over it, pick it up, put it down, pick it up again, and walk around with it a while, weighing it against the other gifts that are wrangling for your regard.

A gift is a gift if you choose it finally, wrap it with one last gust of uncertainty, and give it away. That is a gift. An item, inanimate, that reeks with intention and puts us at risk. And that is why gifts are dangerous and difficult and cause such distress.

What Sean was proposing was another kind of gift — a tribute to each other promoted as a practical trade. He had observed in me a particular passion for books, and he wanted to tell me, in turn, about the satisfaction he had begun to feel in surrounding himself with music. Having just purchased a stereo with a turntable and a tape deck, he wanted not only to own albums but to record them. With his suggestion, he was initiating a pact between us, accelerating our collections, advancing our souls. He was wise that way, and willing, long before I would ever have been bold enough on my own, to buck the system by saying, "If you give me this, I'll give you that."

To retain the spirit of our family's gift-giving, we established some ground rules. We couldn't suggest specific titles; the selection had to be a surprise. We could, however, direct each other to certain sections of a store, so when we were shopping together at the mall a few weeks before Christmas, conducting our pick-up, put-down, is-this-it-or-isn't-it for everyone else on our list, Sean took me into the record store and pointed out the bins of classical music from which he hoped I would choose. He had not heard much classical music — I had little experience with it myself — and his choice was one not of preference (his taste ran to Pink Floyd, Genesis, and Rush) but of the promise of something entirely new. I took him to the bookstore and said, "Anything from the first four shelves," trying to appear arbitrary as we stood in front of a section of paperbacks labeled literature by a black-and-white laminated sign that hung above them. Then we parted and made our purchases privately.

I chose Aaron Copland's *Appalachian Spring* for Sean. I remembered hearing it once in someone else's car. He presented me with Willa Cather's *Death Comes for the Archbishop*. What caused him to choose it, he didn't say. Two weeks later, when he killed himself over an awkward moment of adolescence, he had not yet listened to *Appalachian Spring* all the way through or turned the record over to hear *The Tender Land*.

It took me ten years to read *Death Comes for the Archbishop*. It was the title that stopped me, the words *Death*

Comes. I tried telling myself that there was no connection between Sean's death and the gift he gave me for Christmas. I told myself there was no message, no clue, no cry for help. He was only giving me what I had asked for: anything from the first four shelves. He was giving me Cather, not Death Comes. He was giving me a book, not a warning. Still, something held me back. Each year I took the book from my shelf and looked at the cover. One year I even read the prologue, but I was unable to go any further until finally, a decade after he died, I read *Death Comes for the Archbishop* in three consecutive evenings in January, all of them cold.

So what would I want to tell him? That time passes. Hours, days, and decades. "One summer evening in the year 1848 . . . One afternoon in the autumn of 1851 . . ." Like that. Like chapters ending and opening. I didn't read ahead. I didn't wonder whether death came calmly or suddenly, small or large against the landscape. It was a series of sacraments, an expanse of the sky, each with its own ceremony, none cut short. Isn't it ironic? I would ask him. Isn't it amazing? That death comes for the archbishop at the *end* of his life.

What would I want to tell him? That time passes. That I live in New York now and have friends who never knew him. Two offer me their apartment in Santa Fe every summer, saying, Please, go. It's empty otherwise. I've yet to say okay. Would you want to see the Southwest together, Sean? Would you want to see that part of Willa Cather's country with me — Santa Fe, where death finally comes for the archbishop? I wish you had considered the West, how wide it is. I wish you had wanted, desperately, to see the desert, the mountains, the ocean. No, that is something I learned later, something I've used since. Not fair to ask of you at fifteen what I didn't discover myself until I was three years into my thirties and saw the ocean for the first time. And, as with every new place I've visited without you, I said, "Look, this is what it's like." Thinking, If only you had seen *this,* Sean. This surely would have saved you. This surely would have restored some part of your soul. This New York. This New England. This new part, even, of our own town, all the places in St. Louis that I would have — no, that I should have — shown you.

Regrets? Yes, a world of them. Some nights, lying in your old bed, lying in my own, I think of all the places that you've never been, and I find myself still wanting the world for you. I've seen the ocean, Sean. I've seen mountains. I've seen cities larger than St. Louis. I've seen a world neither of us ever knew. Once, I saw a red jellyfish that had washed up on the shore of an island in the north Pacific. I knelt down next to it, feeling almost humbled by its beauty. It was the first jellyfish I'd ever seen, and as with every new encounter, I had the urge to call out to you. "Sean," I say when no one's around. It was a gray autumn day. The ocean and the sky were nearly the same color. On the beach, the jellyfish looked like a glossy spill, dark maroon at its center, lighter and lighter red toward its edges, the whole shimmering expanse of it covering a wide circle of sand. Every few seconds it heaved and collapsed from its center, sending a ripple out from the dark maroon spot to the paler parts of its body. I knelt there, watching it. I wanted to touch it, but I was afraid. A woman walked by with her dog. "You've never seen a jellyfish?" she asked as the dog sniffed the sand around it. She stood with me for a moment. "If you're thinking you can save it, you can't," she said. She was a native, accustomed to what happened near those waters. I was not. She signaled to her dog, and they continued down the beach. I watched the jellyfish a while longer. I wanted to touch it while it was still living. I knew nothing about jellyfish. Would it sting me? Burn me? These are the things you would know, Sean, the things you could tell me. I reached out and touched its dark maroon middle. When it moved again, sending a single beat through its body, a faint warmth rose up, leaving my hand coated with mucus. I walked to the water and washed it off, and when I returned, the jellyfish was no longer breathing.

Sean, on the same island there were trees with red bark. Below the shedding bark, the trunks were red and orange and yellow. There was something almost arrogant about them, beautiful arrogant trees with bright flesh below bright bark. Flying home, I looked out the window of the plane, and I could see the colors of those trees shouting up at me. Then, within minutes, the scene changed from red and orange and yellow to solid white. You've never flown before. Sometimes, flying, you can see the weather occurring below you. And clouds close up, Sean. This day, all I could see were clouds, hills and hills of them. Until the mountain peak. It broke through abruptly, its jagged whiteness appearing like a sudden authority in the sky. It was wondrous and shocking. I looked around; most of the people on the plane were sleeping. I wanted to wake them. "Sean," I said instead. There, amid all that white softness, a snow-covered mountain. Miles of unbroken clouds and then the mountaintop, an impenetrable peak as white as everything around it. I never want to forget this, I thought, and then, in a brief moment, the mountain was behind us.

Thank you for the Cather. It was a good choice. Thank you for the gift you gave me and for the gift you asked me for. I listened to *Appalachian Spring* in your room after your funeral. Downstairs, Barbra Streisand was singing from the album *Memories*. Mom played that album over and over after you died. I left the door of your room open and sat in the upstairs hallway, listening to *Appalachian Spring* and Barbra Streisand at the same time, getting up only to reset the record. I turned it over to play *The Tender Land*. Downstairs, Mom had *Memories* playing on automatic. I listened to *Appalachian Spring* and *Memories*, *The Tender Land* and *Memories*, *Appalachian Spring*, *The Tender Land*. We had buried you that morning. I sat in the hallway for hours. It grew dark. I couldn't move. I stayed there, listening to Copland and Streisand at the same time, unable to choose one over the other, needing them both, the known and the unknown, the foreign and the familiar. I didn't know where to go or what to do. I lay down in the hallway. The whole second floor of the house was dark. "What are you listening to?" Mary asked when she came up later and turned on the light to say she was leaving. "*The Ten-*

der Land," I told her. She lay down beside me to listen. "You're going to stay here tonight, right?" she said when it was over. "It's snowing," she said, and we lay there, *Memories* rising up from the first floor, until we both fell asleep.

About Cather: after *Death Comes* I read other books she'd written. I read *O Pioneers!* and *My Ántonia* on a train trip from New York to Seattle. I sat in the observation car all day and looked at the landscape. At night, in my sleeping car, I read novels. One night I read Cather's descriptions of the grasslands, and the next day, going through the Great Plains, I saw every word she wrote. The colors I couldn't quite believe would be there were there. The reds and purples and pinks, colors I had never thought of as part of the plains. It was late afternoon. Some of it was the sun, I suppose. Always a skeptic, you used to tell me. And I smiled, thinking of you calling me on it then, seeing the colors and questioning their source.

Why is the sky blue one day and not the next? You had posed that question for your science project one year and brought home an A for your explanation. I remember feeling so proud of you, the simplicity of your approach, your confidence in the face of boys who were building rockets and solar engines and devices that sent off sparks. I remember the cardboard panels you painted that showed the sky in various shades, but I don't remember the answer. Why *is* the sky blue one day and not the next? At your funeral, an old man who lived on the block behind us came up to me. He was someone whose lawn you mowed. He told you once that he was going to a lake somewhere in southern Missouri the following weekend with his brother, and you went back to his house the next day and told him everything you thought he might need to know about that lake — its size, the average depth of the water, its temperature that time of year, what kind of fish were found there. "I'll miss him," the man said after he told me the story. So many people said that, Sean. So many people had stories.

On the train, when I saw the colors Willa Cather wrote about, I had the feeling you were with me. It was autumn, October. I always seem to travel near your birthday, to be coming or going or passing through some place. I was in the observation car. A boy came up and asked me for my autograph. He was nine or ten, wore thick glasses, and stuttered severely. It was his first train trip, he said, and he was collecting the signatures of all the passengers who were on the trip with him. He handed me his autograph book and asked me to write my name and the name of the town where I lived. Without thinking about it, I wrote your name. *Sean Finneran,* I signed. *St. Louis, Missouri.* The boy looked at the book and read aloud what I had written. "Sean. St. Louis, Missouri," he stuttered, not attempting our last name. "Thanks, Sean," he said, and then he approached the man in the seat beside me. Instead of turning back to the landscape, I watched the boy offer his autograph book to everyone around me and I listened as he read aloud each person's entry, all those names and hometowns coming together like a map of America.

Later, walking back to my sleeping car, I passed the boy and his mother sitting in two coach seats by the bathroom. The boy had his book open on his lap, and he read the passengers' names to his mother while she paged through a magazine. That night, after finishing Cather, I lay in my sleeping car, watching the darkness pass over the Great Plains. In the morning, we would be in the mountains. I thought about the boy and his book, how brave he seemed, approaching strangers with his stutter, a nine- or ten-year-old boy on his first train trip. It made me remember the bus trip we took once from our house in the suburbs to downtown St. Louis. In a city of cars, it was your first bus ride. You were twelve, I think; I was twenty-one. The bus was crowded, and we couldn't get seats together. You sat across the aisle from me, next to a man reading the newspaper. I watched you the whole time, your quiet awareness of everyone around you as you took in the parts of the city you had never seen, and I had this sensation that we were the same person, you and I, with the same thoughts and the same sensibilities. Maybe that's why I think I should have seen it coming, Sean. Why didn't I see it coming? Why didn't I ever tell you how close I came myself? Why didn't I think it could happen to you? When I signed your name in the boy's book, why did it take me a few minutes to realize it wasn't mine?

One last thing about Cather, Sean, and Copland — about the last gifts we gave each other. I am spending a few months in the woods of New Hampshire now, at a place where, many years ago, Aaron Copland composed music. Willa Cather is buried nearby. She spent the summers in a town a few miles away, writing in a tent she pitched behind her house. Knowing none of this before I came, I brought *Death Comes for the Archbishop* and *Appalachian Spring* with me. When I was packing them, I wasn't sure why.

This morning, just as I woke up, a sudden rain shower started. I went out on the porch of my cabin and listened to the rain falling through the trees. I wonder if you've ever heard that sound, Sean, the flat, soft suffusion of rain when it falls in the woods. Later, after the rain ended, I went to see the grave of Willa Cather. A pilgrimage, I guess you could call it, though that makes it sound more purposeful — more spiritual — than it was. Still, something drew me to it.

I will never know why you chose the gift you gave me. I will never know why you decided on *Death Comes*, whether it was a chance decision or a choice that meant something more. In the end, it doesn't really matter. Today, as I was reading Willa Cather's headstone, here is what happened: A quote from *My Ántonia* is etched below her name. "That is happiness," it says, "to be dissolved into something complete and great." Maybe it was a memory rising slowly from the rain this morning, one that took all afternoon to reach me. Maybe it was something more. When I read those words — *That is happiness* — I immediately remembered a summer day, long ago, when I was lying on our old couch in the basement reading a book. It was raining, and through the basement windows I could see you in the back yard with your buckets, collecting rainwater for your fish. You were wearing a clear plastic poncho over your T-shirt and shorts. You must have been nine or ten then. It was the summer before I started college. You had just begun breeding guppies. After a while, you tapped on the window. I remember the way your face looked, framed by all that wetness, your blond hair parted by rivulets of rain, your skin glistening. "Will you help me?" you asked when I opened the window. One by one, you lowered your buckets of rainwater into the basement, each time with the same warning: "This one's heavy. This one's really heavy. Watch out for this next one. It's heavy. Careful. Got it?" you'd say. Your buckets safely inside, you came down to the basement and began pouring the water into your aquariums while I lay on the couch, reading. We passed the afternoon that way, neither of us talking, only the sound of water between us, the rain hitting the window, the rain splashing from your buckets into those fifteen glass boxes you kept filled with fish.

That is what I remembered when I read Willa Cather's words: I remembered the day we spent in the basement, I with my book, you with your fish, and I thought, That was happiness. That *is* happiness, Sean, everything dissolved into its simplest, purest form that day; for me, something complete and great. "Will you help me?" you asked, peering in, wet, through the window. "Will you help me?" "Yes," I said. I will. I would.

Back in my cabin now, this morning's rain is a memory. The sky is clear and cloudless. The moon is full. Tonight, I walked in the woods for the first time without a flashlight. I was frightened for a while, but then forgot. The stars were brighter than I'd ever seen. Today is your birthday — yes, it's October, and I am wandering, unreachable, again — but for once, instead of thinking back to the day you were buried, I am remembering the year you were born. One night over dinner, Mom announced that we were getting something new. "A pool," Michael guessed. "A new TV," I tried. "A baby," Mary said, and she was right. It was the first mention of you that I remember. The next morning, at Sunday Mass, a procession of angels appeared in the aisle. I was among them, one of the third-grade girls who got to wear a special gown with wings in the First Communion pageant. In first grade, we made our communion; in second, we were confirmed; and in third, we got to walk down the aisle in the angel outfits we had been waiting with such anticipation to wear. They were costumes made years ago by someone's mother, beautiful satin gowns of all colors — yellow, violet, pink, and blue. I was a pale orange angel. When I walked down the aisle, Mom waved to me and smiled from where she was sitting, Dad, Michael, and Mary beside her, and you there, Mom's special intention, a being barely formed. When I waved back, I felt a brush of wings across my shoulders, like a soft breath behind me. Six months later, on the Feast Day of the Guardian Angels, you arrived in the world, and by the next night, we knew your name.

Sean, time passes, it's true. Hours, days, and decades. And grief goes by its own measure. Now, before this day of angels ends again, before the sky changes color and the moon follows in its phase from full to new, I want to call out your name and tell you, across the tender land, that we have gone on living. We are all, every one of us, alive.